I0055676

DETERMINING THRESHOLDS OF COMPLETE SYNCHRONIZATION, AND APPLICATION

WORLD SCIENTIFIC SERIES ON NONLINEAR SCIENCE

Editor: Leon O. Chua
University of California, Berkeley

*To view the complete list of the published volumes in the series, please visit:
http://www.worldscibooks.com/series/wssnsa_series.shtml

WORLD SCIENTIFIC SERIES ON NONLINEAR SCIENCE

Series Editor: Leon O. Chua

Series A   Vol. 67

DETERMINING THRESHOLDS OF COMPLETE SYNCHRONIZATION, AND APPLICATION

Andrzej Stefański

Technical University of Łódź, Poland

World Scientific

NEW JERSEY · LONDON · SINGAPORE · BEIJING · SHANGHAI · HONG KONG · TAIPEI · CHENNAI

Published by

World Scientific Publishing Co. Pte. Ltd.

5 Toh Tuck Link, Singapore 596224

USA office: 27 Warren Street, Suite 401-402, Hackensack, NJ 07601

UK office: 57 Shelton Street, Covent Garden, London WC2H 9HE

British Library Cataloguing-in-Publication Data
A catalogue record for this book is available from the British Library.

World Scientific Series on Nonlinear Science, Series A — Vol. 67
DETERMINING THRESHOLDS OF COMPLETE SYNCHRONIZATION,
AND APPLICATION

Copyright © 2009 by World Scientific Publishing Co. Pte. Ltd.

All rights reserved. This book, or parts thereof, may not be reproduced in any form or by any means, electronic or mechanical, including photocopying, recording or any information storage and retrieval system now known or to be invented, without written permission from the Publisher.

For photocopying of material in this volume, please pay a copying fee through the Copyright Clearance Center, Inc., 222 Rosewood Drive, Danvers, MA 01923, USA. In this case permission to photocopy is not required from the publisher.

ISBN-13 978-981-283-766-0
ISBN-10 981-283-766-3

Preface

Generally speaking, a phenomenon of synchronization can be defined as a correlation (adjustment) in time of two or more different processes, but it is obvious that in the scientific bibliography of the problem there exist plenty of various definitions and terms describing this phenomenon. Recently, the idea of synchronization took on a more interdisciplinary character, apart from physical and technological applications only, and it has become an object of great interest in many other areas of science. Moreover, this concept has also been adopted for chaotic systems. Earlier, it was supposed that synchronization phenomenon only concerns periodic systems of regular dynamics, while deterministic chaos and synchronization mutually exclude each other due to the sensitivity to the initial state. However, many researchers have demonstrated that two or more chaotic systems can be synchronized by linking them with a mutual coupling or with a common signal.

The results of the above-mentioned research have been described in numerous papers in scientific journals. The authors of larger publications, i.e., books, monographs or surveys, usually give a general overview of the synchronization phenomenon and its possible applications, while this monograph is mainly focused on the complete synchronization problem, i.e., conditions of its occurrence, speaking more precisely. This problem is demonstrated with respect to a type of coupling, which is applied between dynamical systems. Therefore, a detailed classification of such possible couplings is introduced.

Another aspect distinguishing this book among other publications on synchronization is an application of the synchronization properties for the estimation of Lyapunov exponents, especially for non-differentiable systems. As is well known, these exponents are one of the most sophisticated tools to identify a character of motion of dynamical systems and their calculation or estimation is one of the fundamental tasks in studies of these systems. For practical applications, it is enough to know the largest Lyapunov exponent. If it is positive, then the system is chaotic. A non-positive maximum number indicates regular system dynamics. On the other hand, in the experimental practice a regular motion is manifested by the synchronization phenomenon. In turn, a lack of synchronization indicates irregular dynamics of chaotic or stochastic nature. Thus, there appears an explicit correlation between synchronous/ desynchronous states and values of Lyapunov exponents. Hence, the synchronization process can be applied to detect chaos and also to determine the largest Lyapunov exponent of an arbitrary dynamical system.

Generally, the proposed monograph is composed of two parts, where:
1. modern techniques for determining the synchronization thresholds,
2. application of the complete synchronization for the estimation of Lyapunov exponents, are analyzed and described.

In the first part (Chapters 1 – 4), after a general introduction to the synchronization phenomenon, a classification of couplings between dynamical systems is presented. Next, a review of analytical and numerical methods (e.g., a graph method, a concept called master stability function) to determine the synchronization thresholds for identical or slightly different dynamical systems is demonstrated and illustrated with several examples of the oscillatory networks with single or disconnected synchronous range(s) of the parameter space. The second part of the book (Chapters 5 – 7) is devoted to a description of the synchronization method for the estimation of the largest Lyapunov exponent with a review of its possible applications. This depiction is preceded with a short survey of classical and modern techniques for the determination of Lyapunov exponents. The properties of the diagonal coupling considered in the first part of the book are the basis for the proposed technique. The main advantage of this approach lies in its usefulness for dynamical systems with discontinuities or time delay, where classical attempts are not easily applicable. Therefore, the proposed method is mainly exemplified with cases of such systems.

February 2009 *A. Stefanski*
Lodz

Contents

PART I
THEORY

Chapter 1

Introduction

1.1 General

Collective motion of dynamical systems has been known for a long time, i.e., since the second half of 17th century, when Huygens discovered the synchronization of two clock pendulums (Huygens (1673)). Next, this phenomenon has been observed and investigated in various types of mechanical or electrical systems (Rayleigh (1945), Van der Pol (1920), Blekhman (1920)). In theory and practical analysis of dynamical systems, the research of an interaction between them plays an important role. Such an interaction often leads to an appearance of some synchronization effects. Especially interesting are oscillators exhibiting chaotic or stochastic dynamics. In recent years, the chaotic synchronization has become an object of great interest in many areas of science, e.g., biology (Hertz *et al.* (1991), Soen *et al.* (1999)), communication (Coumo *et al.* (1993)) or laser physics (Winful & Rahman (1990), Liu, Rios Leite (1994)). There have appeared some new interesting ideas concerning oscillatory networks, e.g., a concept of the so-called small-world networks (Watts (1999)), which include the properties of regular and random networks, or the scale-free property (Albert *et al.* (1999)), which is signified by the power-law connectivity distribution of the network. Over the last decade, a number of new types of synchronization have been also identified, e.g., generalized synchronization (Rulkov (1995); Kocarev, Parlitz (1996)), phase (Rosenblum *et al.* (1996), Pikovsky *et al.* (1997)), anti-phase (Cao, Lai

(1998)) and imperfect phase synchronization (Zaks *et al.* (1999)), lag (Rosenblum *et al.* (1997)) and anticipated synchronization (Voss (2000), (2001); Masoller (2001)).

In general, the dynamics of any set (network) of N interacting oscillators can be described in the following block form:

$$\dot{\mathbf{x}} = \mathbf{F}(\mathbf{x}) + \sum_{j=1}^{M} \sigma[\mathbf{G}_j \otimes \mathbf{H}_j(\mathbf{x})], \qquad (1.1a)$$

for flows, and

$$\mathbf{x}_{n+1} = \mathbf{F}(\mathbf{x}_n) + \sum_{j=1}^{M} \sigma[\mathbf{G}_j \otimes \mathbf{H}_j(\mathbf{x}_n)], \qquad (1.1b)$$

for maps. Here $\mathbf{x} = (\mathbf{x}_1, \mathbf{x}_2, \ldots, \mathbf{x}_N) \in \Re^m$, $\mathbf{F}(\mathbf{x}) = (\mathbf{f}_1(\mathbf{x}_1), \ldots, \mathbf{f}_N(\mathbf{x}_N))$, $\mathbf{H}_j : \Re^m \to \Re^m$ are linking (output) functions of each oscillator variables that are used in the coupling, $\mathbf{G}_j$ is the connectivity matrix, i.e., the Laplacian matrix representing the M-number of possible topologies of connections between the network nodes corresponding to a given linking function $\mathbf{H}_j$, σ is an overall coupling coefficient and $\otimes$ is a direct (Kronecker) product of two matrices (Barnett, Storey (1970)). Such a product of two matrices $\mathbf{G}$ and $\mathbf{H}$ is given in the block form by:

$$\mathbf{G} \otimes \mathbf{H} = \begin{pmatrix} G_{11}\mathbf{H} & G_{12}\mathbf{H} & \cdots & G_{1N}\mathbf{H} \\ G_{21}\mathbf{H} & G_{22}\mathbf{H} & \cdots & G_{2N}\mathbf{H} \\ \vdots & \vdots & \ddots & \vdots \\ G_{N1}\mathbf{H} & G_{N2}\mathbf{H} & \cdots & G_{NN}\mathbf{H} \end{pmatrix}. \qquad (1.2)$$

The above Eqs. (1.1a) and (1.1b) describe a general case of the oscillatory network, where there are different (i.e., they can be non-identical) m-dimesional node systems $\mathbf{f}_i(\mathbf{x}_i)$ with an arbitrary topology of connections and different linking functions $\mathbf{H}_j$. However, further considerations are restricted to the case of identical node systems and linking functions because, as has been mentioned in Preface, this monograph deals with the phenomenon of the so-called *complete synchronization* (explained below) and its practical relation to the stability theory and Lyapunov exponents.

1.2 Complete Synchronization

Pecora and Carroll (1990) have defined the *complete synchronization* (CS) as a state when two state trajectories $\mathbf{x}(t)$ and $\mathbf{y}(t)$ converge to the same values and continue in such a relation further in time. This phenomenon takes place between two identical dynamical systems $\dot{\mathbf{x}} = \mathbf{f}(\mathbf{x})$ and $\dot{\mathbf{y}} = \mathbf{f}(\mathbf{y})$ (when separated). If some kind of linking between them is introduced (a direct diffusive or inertial coupling, a common external signal, *etc.*), the CS, i.e., full coincidence of phases (frequencies) and amplitudes of their responses, becomes possible.

Definition 1.1 *The complete synchronization of two dynamical systems represented with their phase plane trajectories* $\mathbf{x}(t)$ *and* $\mathbf{y}(t)$, *respectively, takes place when for all t > 0, the following relation is fulfilled:*

$$\lim_{t\to\infty}\left\|\mathbf{x}(t) - \mathbf{y}(t)\right\| = 0. \tag{1.3}$$

It is also described in the subject literature as the identical or full synchronization (Pecora and Carroll (1990), Rosenblum *et al.* (1997)).

1.3 Imperfect Complete Synchronization

The CS state can be reached only when two identical dynamical systems are concerned, say, they are given with the same ODEs with identical system parameters. This condition of identity may not be fulfilled due to presence of an external noise or parameters mismatch, which usually can happen in real systems. If the scale of such disturbances is relatively small, then both systems may eventually reach a state called the *imperfect complete synchronization* (ICS), sometimes referred to as the practical or disturbed synchronization (Liu, Rios Leite (1994), Kapitaniak *et. al.* (1996), Sekieta, Kapitaniak (1996)).

Definition 1.2 *The imperfect complete synchronization of two dynamical systems represented with their phase plane trajectories* $\mathbf{x}(t)$ *and* $\mathbf{y}(t)$, *respectively, occurs when for all t > 0, the following inequality is fulfilled:*

$$\lim_{t \to \infty} \|\mathbf{x}(t) - \mathbf{y}(t)\| < \varepsilon, \qquad (1.4)$$

where ε is a small parameter, such that $\varepsilon \ll \sup\|\mathbf{x}(t) - \mathbf{y}(t)\|$.

1.4 Generalized Synchronization

One of the most interesting ideas concerning the chaos synchronization, which have emerged in the last years, is a concept called the *generalized synchronization* (GS). This term has been introduced by Rulkov *et al.* (1995) as a generalization of the synchronization idea for unidirectionally coupled systems:

$$\dot{\mathbf{x}} = \mathbf{f}(\mathbf{x}), \qquad (1.5a)$$

$$\dot{\mathbf{y}} = \mathbf{g}(\mathbf{y}, \mathbf{h}(\mathbf{x})) \qquad (1.5b)$$

where $\mathbf{x} \in \Re^m$, $\mathbf{y} \in \Re^k$ and $\mathbf{h}(\mathbf{x})$: $\Re^m \to \Re^k$ is a function characterizing the coupling between the drive (Eq. (1.5a)) and the response (Eq. (1.5b)) system. Such a kind of interaction of dynamical systems is also called the *master–slave coupling*. We can say that the GS of these systems occurs if there exists a static functional relation ψ between their states, i.e.,

$$\mathbf{y}(t) = \mathbf{\Psi}[\mathbf{x}(t)]. \qquad (1.6)$$

Generally, the GS problems have been researched both in the context of identical (when separated) systems (1.5a) and (1.5b), and also in cases when the response system (the same set of ODEs with different values of system parameters) is slightly or strictly different (another set of ODEs) than the driving oscillator (Abarbanel *et al.* (1996), Kocarev, Parlitz (1996), Pyragas (1996), Boccaletti *et al.* (2002)). The GS phenomena can be also observed in discrete time systems (Pyragas (1998), Afraimovich *et al.* (2002)). However, in any of these cases the CS can be applied as a tool for recognizing the GS. In order to detect the presence of the GS, a numerical method called the *mutual false nearest neighbors* (Rulkov *et al.* (1995)) and the related *auxiliary system approach* (Abarbanel *et al.* (1996)) have been proposed. According to these methods, the criterion for the GS existence is an appearance of the CS between the response subsystem (Eq. (1.5b)) and its identical replica, i.e.,

$$\lim_{t \to \infty} \left\| \mathbf{y}(t, \mathbf{x}_0, \mathbf{y}_{01}) - \mathbf{y}(t, \mathbf{x}_0, \mathbf{y}_{02}) \right\| = 0 \tag{1.7}$$

where $(\mathbf{x}_0, \mathbf{y}_{01})$ and $(\mathbf{x}_0, \mathbf{y}_{02})$ are two generic initial conditions of systems (1.5a) and (1.5b). An occurrence of such CS (Eq. (1.7)) indicates that slave systems forget their initial states, so their functional control defined by Eq. (1.6) takes place.

The properties of the synchronization manifold allow us to divide the GS into two types (Pyragas (1998)):

1. The weak GS, which takes place for the continuous but non-smooth map ψ when the global dimension of the strange attractor d^G, located in the whole phase space $X \oplus Y$, is larger than the attractor dimension of the driving system d^D, i.e.,

$$d^G > d^D. \tag{1.8}$$

2. The strong GS, when the functional ψ is smooth and we have:

$$d^G = d^D, \tag{1.9}$$

i.e., the response oscillator does not influence the global attractor.

The attractor dimensions d^G and d^D can be estimated on the basis of the Lyapunov exponents spectrum according to the Kaplan & Yorke conjecture (Kaplan & Yorke (1979)) defined by the formula (5.19) in Chapter 5.

The properties and applications of the weak and strong GS are described in more detail in Chapter 4 (Sec. 4.3) in the context of the synchronizability of externally driven oscillators.

Chapter 2

Classification of Couplings

In the literature dealing with networks of coupled oscillators or multi-degree-of-freedom systems, a huge number of definitions and terms describing various kinds of couplings between the dynamical systems can be found (Pecora & Carroll (1990), (1991), Watts & Strogatz (1998), Watts (1999), Albert *et al.* (1999), Kocarev & Parlitz (1995), Boccaletti *et al.* (2002), Hale (1997), Park *et al.* (2008), Pikovsky *et al.* (2001)). Consequently, in these works various criterions of the coupling classification are given (Chua (1994)). The classification of couplings presented here can be treated as a part of the theoretical background for the investigation of the CS phenomenon. Therefore, this survey is mainly focused on cases making an occurrence of such synchronization possible. A decisive factor for distinguishing coupling schemes is a structure of the linking function $\mathbf{H}$ or the connectivity matrix $\mathbf{G}$ (Eqs. (1.1a–b)).

2.1 Negative Feedback

Negative feedback is a process commonly met in nature and human environment, i.e., in many physical, biological, chemical and engineering systems. This idea can be also applied to explain some social or economical mechanisms. *Negative feedback* causes that a part of the system output, inverted, is supplied into the system input; generally, as a result, fluctuations are damped (their amplitude decreases). Many real-world systems have one or several points around which the system oscillates. In response to a perturbation, a negative feedback system with such point(s) will tend to re-establish the equilibrium. In engineering, mathematics and physical and biological sciences, common terms for the points around which the system gravitates include: attractors, stable states, equilibrium points. The term *negative* refers to the sign of the multiplier in mathematical models for the feedback.

In contrast, *positive feedback* is a feedback in which the system responds in the same direction as the perturbation, resulting in amplification of the original signal instead of stabilizing it.

Negative feedback is a good tool for control and self-regulation of dynamical systems. It is also the most common method that leads the synchronization of coupled oscillators. In general, an *i*-th subsystem connected to other *N*–1 subsystems of the entire system can be represented as:

$$\dot{\mathbf{x}}_i = \mathbf{f}(\mathbf{x}_i) + \sum_{j=1}^{N-1} \sigma G_{ij} \mathbf{D}\mathbf{g}(\mathbf{x}_i, \mathbf{x}_j), \qquad (2.1)$$

where $\mathbf{g}(\mathbf{x}_i, \mathbf{x}_j) \in \Re^m$ is a coupling vector and $\mathbf{D}$ is an *m*✕*m* linking matrix of constant components. Both $\mathbf{D}$ and $\mathbf{g}$ are the same for each pair of coupled oscillators. Thus, according to Eqs. (1.1a–b), the output function is $\mathbf{H}(\mathbf{x}) = \mathbf{D}\mathbf{g}(\mathbf{x}_i, \mathbf{x}_j)$. The map version of the *negative feedback coupling* is analogous to the time-continuous case given by Eq. (2.1). Various schemes and kinds of the *negative feedback coupling* can be classified, as presented below, with different forms of the coupling vector $\mathbf{g}$, the connectivity matrix $\mathbf{G}$ and the output function $\mathbf{H}$.

2.1.1 *Linear and nonlinear coupling*

The coupling between dynamical systems is called *linear coupling* if all components of the coupling vector $\mathbf{g}$ (Eq. (2.1)) are linear terms, i.e.,

$$\mathbf{g}(\mathbf{x}_i, \mathbf{x}_j) = [x_{j1} - x_{i1}, x_{j2} - x_{i2}, ..., x_{jm} - x_{im}]^{\mathrm{T}}. \qquad (2.2)$$

On the other hand, if components of the coupling vector are not defined linearly, then the coupling is *nonlinear*, e.g.,

$$\mathbf{g}(\mathbf{x}_i, \mathbf{x}_j) = [(x_{j1} - x_{i1})^3, (x_{j2} - x_{i2})^3, ..., (x_{jm} - x_{im})^3]^{\mathrm{T}}. \qquad (2.3)$$

The examples presented in next Chapters of this book are the cases of the *linear negative feedback*.

2.1.2 *Mutual and unidirectional coupling*

Consider the simplest, double-oscillators case ($N=2$) of the system (1.1a) with the *negative feedback coupling*:

$$\begin{Bmatrix} \dot{\mathbf{x}}_1 \\ \dot{\mathbf{x}}_2 \end{Bmatrix} = \begin{Bmatrix} \mathbf{f}_1(\mathbf{x}_1) \\ \mathbf{f}_2(\mathbf{x}_2) \end{Bmatrix} + \sigma \begin{pmatrix} -\mathbf{D}_1 & \mathbf{D}_1 \\ \mathbf{D}_2 & -\mathbf{D}_2 \end{pmatrix} \begin{Bmatrix} \mathbf{x}_1 \\ \mathbf{x}_2 \end{Bmatrix}, \tag{2.4}$$

which is schematically depicted in Fig. 2.1. The corresponding connectivity matrix is:

$$\mathbf{G} = \begin{pmatrix} -1 & 1 \\ 1 & -1 \end{pmatrix}. \tag{2.5}$$

The connection of subsystems in system (2.4) is called the *mutual coupling* due to a bidirectional interaction between both subsystems, which causes reciprocal control of each other behavior. For $\mathbf{D}_1 = \mathbf{D}_2$, a *symmetrical mutual coupling* takes place.

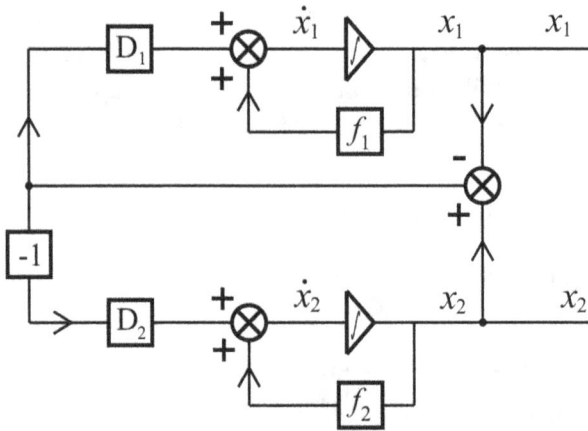

Fig. 2.1. A scheme the *negative feedback coupling* of two systems.

When one of the linking matrices in Eq. (2.4) possesses all zero elements (say $\mathbf{D}_1 = \mathbf{0}$), the coupling is *unidirectional*. Then, the dynamics of one of the coupled systems (i.e., $\dot{\mathbf{x}}_1 = \mathbf{f}_1(\mathbf{x}_1)$) is independent of the coupling because the related connectivity matrix has a form:

$$\mathbf{G} = \begin{pmatrix} 0 & 0 \\ 1 & -1 \end{pmatrix}. \tag{2.6}$$

Thus, in this scheme of the coupling, the dominant (reference) system controls the behavior of its disturbed neighbor. Therefore, such a configuration is called a *drive–response* or *master–slave coupling*.

2.1.3 Diffusive, global and hybrid coupling

The previous Sec.2.1.2 has shown that the coupling can be classified by a form of the matrix $\mathbf{G}$. The structure of the connectivity is a good property for categorizing the type of coupling in larger populations (networks) of oscillators.

One of well-understood coupling mechanisms in such networks is the *global coupling* (also called the *all-to-all coupling*) through which each oscillator interacts with equal strength with all of the other oscillators in the system (Fig. 2.2a). Hence, the matrix $\mathbf{G}$ has a regular symmetrical structure:

$$\mathbf{G} = \begin{bmatrix} 1-N & 1 & \cdots & 1 \\ 1 & \ddots & \ddots & \vdots \\ \vdots & \ddots & \ddots & 1 \\ 1 & \cdots & 1 & 1-N \end{bmatrix}, \tag{2.7}$$

and the dynamics of a single node of such a network with the *linear coupling* is described by:

$$\dot{\mathbf{x}}_i = \mathbf{f}(\mathbf{x}_i) + \sum_{j=1}^{N} G_{ij}\mathbf{D}(\mathbf{x}_j - \mathbf{x}_i). \tag{2.8}$$

On the other hand, the *diffusive coupling* is a qualitatively different scheme of interactions in the network due to its local character. Such a connection is equally referred to as the *nearest-neighbor coupling* (Fig. 2.2b). For an arbitrary *i*-th oscillator, we have:

$$\dot{\mathbf{x}}_i = \mathbf{f}(\mathbf{x}_i) + \mathbf{D}(\mathbf{x}_{j-1} + \mathbf{x}_{j+1} - 2\mathbf{x}_i), \tag{2.9}$$

and the corresponding connectivity matrix is also symmetrical:

$$
\mathbf{G} = \begin{bmatrix}
-2 & 1 & 0 & \cdots & 0 & 1 \\
1 & -2 & 1 & \ddots & \cdots & 0 \\
0 & 1 & \ddots & \ddots & \ddots & \vdots \\
\vdots & \ddots & \ddots & \ddots & 1 & 0 \\
0 & \cdots & \ddots & 1 & -2 & 1 \\
1 & 0 & \cdots & 0 & 1 & -2
\end{bmatrix}. \tag{2.10}
$$

The *diffusive coupling* has been initially introduced on the basis of a diffusion-like process (Bar-Eli (1985)). Unlike the *global coupling*, which generates a mean field in the ensemble of oscillators, the *diffusive coupling* produces a local interaction only between each component of the network and its nearest neighbors (Park *et al.* (2008)).

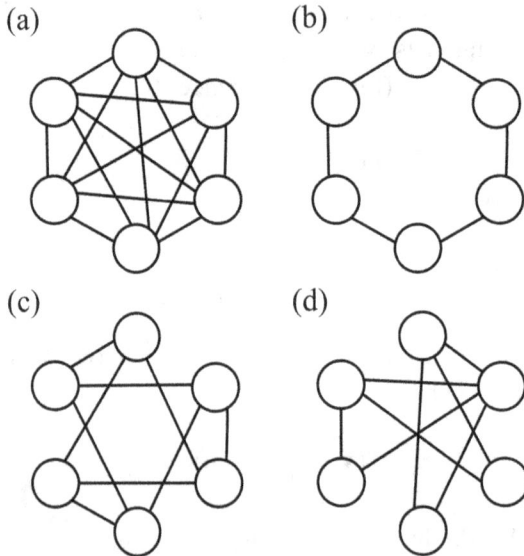

Fig. 2.2. Types of network connections: (a) *global (all-to-all) coupling*, (b) *diffusive (nearest neighbor) coupling*, regular hybrid coupling and (d) *random coupling*.

In general, it is easy to imagine the networks, which are an intermediate case between the *diffusive* and *global couplings* (see Fig. 2.2c). Such hybrids are the examples of regular sets of coupled oscillators as well, because the dynamics of each node is described by the same ODE, in analogy to Eqs. (2.8) and (2.9). Moreover, a star coupling configuration (see Figs. 2.4 or 4.1), where one distinguished central node is coupled to all remaining oscillators in the network, can be also consider as a regular network.

On the other hand, the topology of connections in the network can be completely random (see Fig. 2.2d). Recently, a very interesting concept called small-world networks, which includes the properties of regular and random networks, has been proposed (Watts & Strogatz (1998), Watts (1999)). It has been shown that an addition of a few long-range shortcuts to a diffusively coupled set of oscillators significantly reduces the average distance between nodes, while the entire network is still relatively localized (Barahona, Pecora (2002)).

2.1.4 *Real and imaginary coupling*

The structure of the connectivity matrix is not a sole factor that can be used for the coupling classification. They can be classified also on the basis of eigenvalues γ_j ($j = 0, 1, 2, \ldots, N-1$) of the matrix **G**. According to the *Master Stability Function* (MSF) concept (Pecora & Carroll (1998), Fink *et al.* (2000)), described in detail in the next Sec., the synchronizability of a network of oscillators can be quantified by the eigenvalue spectrum of the connectivity matrix. For an arbitrary configuration of connections between network nodes, all or a part of these eigenvalues can be complex numbers, i.e.,

$$\gamma_j = \alpha_j + i\beta_j, \tag{2.11}$$

where α_j and β_j are real and imaginary components of the eigenvalue, respectively.

Let us consider two extremely opposite variants, i.e., only real (i) or only imaginary (ii) eigenvalues of the matrix **G**.

2.1.4.1 *Real coupling*

If the coupling between the oscillators is mutual and symmetrical, then it results in the symmetric connectivity matrix **G**. Such a matrix possesses only real eigenvalues, so $\alpha_j \neq 0$ and $\beta_j = 0$. Hence, this symmetric coupling is called the *real coupling* and can be interpreted as a kind of damping (Fink *et al.* (2000)). Such a situation takes place in coupled mechanical systems, where an interaction is mutual. The instances of the *real coupling* are *global* and *diffusive couplings* represented by symmetric matrices in Eq. (2.7) and Eq. (2.10), respectively. All the real eigenvalues of them can be calculated according to following analytical formulas:

$$\gamma_j = -N \tag{2.12a}$$

for the *global*, and

$$\gamma_j = -4\sin^2 \frac{j\pi}{N} \tag{2.12b}$$

for *diffusive coupling*.

2.1.4.2 *Imaginary coupling*

On the other hand, a "pure" *imaginary coupling*, i.e., only imaginary eigenvalues can be found in the spectrum ($\alpha_j = 0$ and $\beta_j \neq 0$), occurs if the connectivity matrix is fully anti-symmetric. In order to exemplify this case, let us consider three dynamical systems coupled as follows:

$$\begin{aligned}
\dot{\mathbf{x}}_1 &= \mathbf{f}(\mathbf{x}_1) + \sigma\, \mathbf{D}(\mathbf{x}_3 - \mathbf{x}_2), \\
\dot{\mathbf{x}}_2 &= \mathbf{f}(\mathbf{x}_2) + \sigma\, \mathbf{D}(\mathbf{x}_1 - \mathbf{x}_3), \\
\dot{\mathbf{x}}_3 &= \mathbf{f}(\mathbf{x}_3) + \sigma\, \mathbf{D}(\mathbf{x}_2 - \mathbf{x}_1).
\end{aligned} \tag{2.13}$$

The corresponding connectivity matrix is anti-symmetric with zero components on the diagonal:

$$\mathbf{G} = \begin{bmatrix} 0 & -1 & 1 \\ 1 & 0 & -1 \\ -1 & 1 & 0 \end{bmatrix}, \tag{2.14}$$

and two imaginary non-zero eigenvalues: $\gamma_1 = -i\sqrt{3}$, $\gamma_2 = i\sqrt{3}$. Thus, we can see that the term *imaginary coupling* comes from anti-symmetric connections of oscillators and the interpretation of it is not a damping, but a rotation between two eigenmodes.

2.1.5 Diagonal and non-diagonal coupling

On the contrary to the cases presented above, the *diagonal* or *non-diagonal coupling* is qualified with the structure of the linking function **H**. The exemplary cases of the output function for a 3-D dynamical system, e.g., a Rössler circuit (Rössler (1976)) or a Lorenz oscillator (Lorenz (1963)) are as follows:

$$\mathbf{H} = \begin{pmatrix} 1 & 0 & 0 \\ 0 & 1 & 0 \\ 0 & 0 & 1 \end{pmatrix}, \tag{2.15a}$$

$$\mathbf{H} = \begin{pmatrix} 1 & 0 & 0 \\ 0 & 1 & 0 \\ 0 & 0 & 0 \end{pmatrix}, \tag{2.15b}$$

$$\mathbf{H} = \begin{pmatrix} 0 & 0 & 0 \\ 1 & 0 & 0 \\ 0 & 0 & 0 \end{pmatrix}, \tag{2.15c}$$

$$\mathbf{H} = \begin{pmatrix} 1 & 0 & 0 \\ 1 & 1 & 0 \\ 0 & 0 & 0 \end{pmatrix}, \tag{2.15d}$$

$$\mathbf{H} = \begin{pmatrix} 1 & 0 & 0 \\ 1 & 1 & 0 \\ 0 & 0 & 1 \end{pmatrix}. \tag{2.15e}$$

A *complete diagonal* (CD) *coupling* is realized by all diagonal components of the output function (see Eq. (2.15a)) for each pair of coupled subsystems. Eq. (2.15b) illustrates a case of a *partly diagonal (PD) coupling*, which is realized by not all diagonal components of the output function. If all the diagonal components of the linking function are equal to zero, then the coupling is non-diagonal (ND) — see Eq. (2.15c). Obviously, combined schemes of the coupling are also possible, i.e., a PD–ND configuration (Eq. (2.15d)) or a CD–ND coupling (Eq. (2.15e)).

2.1.6 Dissipative, conservative and inertial coupling

Such possibilities of *diffusive couplings* are in fact counterparts of the *diagonal* and *non-diagonal coupling* cases described above and these terms can be used interchangeably. In order to demonstrate this equivalence of the terms, let us consider simple examples of the coupled mechanical systems depicted in Figs. 2.3. The motion of each mass in the double-oscillatory system from Fig. 2.3a is described by the following pair of ODEs:

$$\dot{x}_i = y_i,$$
$$\dot{y}_i = -x_i - hy_i + \sigma_c(x_j - x_i) + \sigma_d(y_j - y_i), \tag{2.16}$$

where $i, j = 1,2$. The corresponding linking matrix is:

$$\mathbf{H} = \begin{pmatrix} 0 & 0 \\ \sigma_c & \sigma_d \end{pmatrix}. \tag{2.17}$$

Thus, for $\sigma_c = 0$ and $\sigma_d > 0$, the coupling is of the PD type. Such a coupling has a dissipative character because it is realized by the damping component (viscous damper) with dissipation proportional to the velocity y. Therefore, it is also called the *velocity coupling* (Hale (1997)). A *dissipative linear coupling* always results in a decrease of the total divergence of the system which is reduced to the value "$-\sigma_d$". In the opposite case ($\sigma_c > 0$ and $\sigma_d = 0$), the ND coupling is applied to system (2.16). This *position coupling*, which uses the displacement x (Hale (1997)), is conservative because it does not contribute in any way to dissipation, so the divergence of the system does not change in the presence of the coupling. If both coupling coefficients are nonzero, then we have a combined *conservative–dissipative coupling* of the PD–ND type.

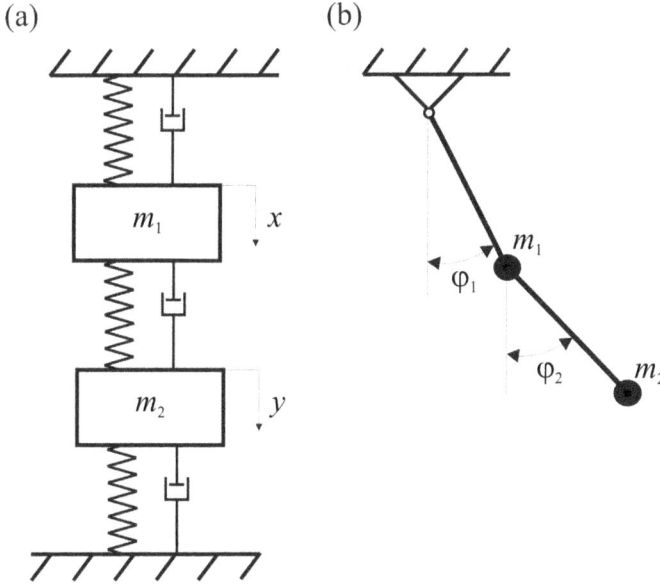

Fig. 2.3. 2-DoF coupled mechanical systems: (a) Masses-springs-dampers connection, (b) double pendulum.

In Fig. 2.3b an *inertial coupling* of two mathematical pendulums is shown. Here, a cause of linking are inertial components of differential equations. The linear version of the double pendulum from Fig. 2.3b is given in the form:

$$\ddot{\varphi}_1 + 0.5\ddot{\varphi}_2 + \omega^2\varphi_1 = 0,$$
$$\ddot{\varphi}_1 + \ddot{\varphi}_2 + \omega^2\varphi_1 = 0,$$

(2.18)

where $\omega^2 = g/l$. It is clearly visible (Eqs. (2.18)) that both pendulums are coupled with inertial variables, so we can refer to it as the *acceleration coupling*. However, the 2nd order ODEs (Eqs. (2.18)) can be easily transformed into a set of the 1st order ODEs with a non-symmetrical *conservative* (i.e., the ND type) *coupling*:

$$\dot{x}_1 = x_2,$$
$$\dot{x}_2 = -\omega^2 x_1 + \omega^2(y_1 - x_1),$$
$$\dot{y}_1 = y_2,$$
$$\dot{y}_1 = 2\omega^2(x_1 - y_1),$$

(2.19)

where $\varphi_1 = x_1$, $\varphi_2 = y_1$.

To summarize these considerations, it can be stated that the terms *diagonal* or *non-diagonal couplings* on one hand and *conservative* or *dissipative couplings* on the other hand can mutually complement each other. First of them seems to be more appropriate from the mathematical point of view, while the second one refers to real physical cases, i.e., mechanical or electrical systems.

An idea of the dissipative and *inertial coupling* has been also adapted for discrete-time systems[1]. Consider two maps $\mathbf{x}_{n+1} = \mathbf{f}(\mathbf{x}_n)$ and $\mathbf{y}_{n+1} = \mathbf{f}(\mathbf{y}_n)$ coupled according to the following formula:

$$\mathbf{x}_{n+1} = \mathbf{f}(\mathbf{x}_n) + \sigma[\mathbf{f}(\mathbf{y}_n) - \mathbf{f}(\mathbf{x}_n)],$$
$$\mathbf{y}_{n+1} = \mathbf{f}(\mathbf{y}_n) + \sigma[\mathbf{f}(\mathbf{x}_n) - \mathbf{f}(\mathbf{y}_n)].$$

(2.20)

Such a coupling tends to equalize the instantaneous states of coupled systems, thus it may be called the *dissipative coupling*. The same maps can be also linked in a different way:

$$\mathbf{x}_{n+1} = \mathbf{f}(\mathbf{x}_n) + \sigma(\mathbf{y}_n - \mathbf{x}_n),$$
$$\mathbf{y}_{n+1} = \mathbf{f}(\mathbf{y}_n) + \sigma(\mathbf{x}_n - \mathbf{y}_n).$$

(2.21)

Then, the coupling keeps memory of the state from the previous step of evolution, and it may be referred to naturally as the *inertial coupling*.

2.2 Drive with a Common Signal

Another kind of coupling which can lead to the synchronous behavior of dynamical systems is a drive with a common signal. In order to demonstrate such a case, let us consider an array of N identical oscillators driven by the common external excitation $\mathbf{e}(t)$ for flows or $\mathbf{e}_n$ for maps. There is not any kind of direct linking between them (a

---

[1] www.sgtnd.tserv.ru/science

negative feedback, a *diffusive* or *inertial coupling*, *etc.*). Thus, this case can be reflected as a star-type connection between oscillators, where a *unidirectional coupling* from the central node (exciter) to the remaining oscillators is realized, as shown schematically in Fig. 2.4. The dynamics of the entire system can be described in the block form:

$$\dot{\mathbf{X}} = \mathbf{F}(\mathbf{X}) + q\big(\mathbf{1}_N \otimes \mathbf{h}(\mathbf{e})\big) \tag{2.22}$$

for flows, and

$$\mathbf{X}_{n+1} = \mathbf{F}(\mathbf{X}_n) + q\big(\mathbf{1}_N \otimes \mathbf{h}(\mathbf{e}_n)\big) \tag{2.23}$$

for maps.

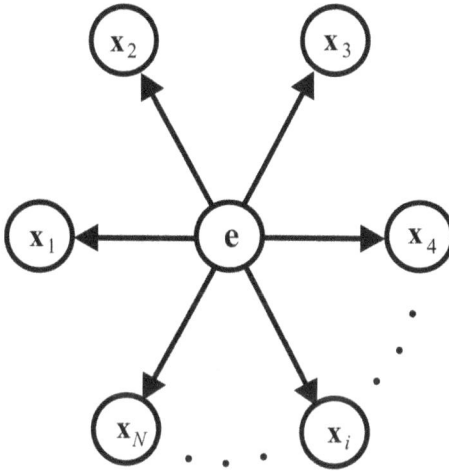

Fig. 2.4. Common drive — a star configuration with a unidirectional coupling.

Here $\mathbf{X} = [\mathbf{x}_1, \mathbf{x}_2, \dots, \mathbf{x}_N]^T$, $\mathbf{x}_i \in \mathfrak{R}^m$, $\mathbf{F}(\mathbf{X}) = [\mathbf{f}(\mathbf{x}_1), \dots, \mathbf{f}(\mathbf{x}_N)]^T$, $\mathbf{1}_N$ is the $N \times N$ identity matrix, q is the overall driving strength, $\otimes$ is a direct (Kronecker) product of two matrices, and $\mathbf{h}: \mathfrak{R}^m \to \mathfrak{R}^m$ is an output function of the external excitation variables $\mathbf{e} = [e_1, e_2, \dots, e_k]^T \in \mathfrak{R}^k$ that are used in the drive.

In order to recognize the synchronizability of response oscillators (Eqs. (2.22) and (2.23)), the properties of the GS have been employed (see Sec. 3.3.3). A real equivalent of commonly driven oscillators is shown in Fig. 4.31.

2.3 Autonomous Driver Decomposition

An idea of *autonomous driver decomposition* was introduced by Pecora & Carroll (1990) as a one of the first coupling configurations making the chaos synchronization possible. In order to explain this issue, let us consider an autonomous k-dimensional ($\mathbf{z} \in \mathfrak{R}^k$) dynamical system:

$$\dot{\mathbf{z}} = \mathbf{f}(\mathbf{z}). \tag{2.24}$$

The system (2.24) can be arbitrarily divided into two subsystems. One of them is m-dimensional, thus the second one is $(k-m)$-dimensional, i.e.,

$$\dot{\mathbf{y}} = \mathbf{g}(\mathbf{y}, \mathbf{w}), \tag{2.25a}$$

$$\dot{\mathbf{w}} = \mathbf{h}(\mathbf{y}, \mathbf{w}). \tag{2.25b}$$

where $\mathbf{y} = [z_1, z_2, \dots, z_m]^T$, $\mathbf{g} = [f_1(z), f_2(z), \dots, f_m(z)]^T$, $\mathbf{w} = [z_{m+1}, z_{m+2}, \dots, z_k]^T$ and $\mathbf{h} = [f_{m+1}(z), f_{m+2}(z), \dots, f_k(z)]^T$.

Create now a new subsystem $\mathbf{w}'$ identical to the $\mathbf{w}$ subsystem:

$$\dot{\mathbf{w}}' = \mathbf{h}(\mathbf{y}, \mathbf{w}'). \tag{2.26}$$

As a result, we obtained an augmented $(2k{-}m)$-dimensional system. Here, Eqs. (2.25a) and (2.25b) define the driving system, whereas the system of Eq. (2.26) represents the response subsystem. Its evolution is controlled by a driving signal $\mathbf{y}$ from subsystem (2.25a).

The concept of *autonomous driver decomposition* can be illustrated more clearly by substituting, e.g., a classical Rössler system into general Eqs. (2.25a–b) and (2.26):

$$
\text{drive}
\begin{cases}
\dot{\mathbf{y}} = \mathbf{g}(\mathbf{y}, \mathbf{w}) & \begin{cases} \dot{x} = -y - z, \\ \dot{y} = x + ay, \\ \dot{z} = b + z(x - c), \end{cases} \\[6pt]
\dot{\mathbf{w}} = \mathbf{h}(\mathbf{y}, \mathbf{w})
\end{cases}
$$

$$
\text{response}
\begin{cases}
\dot{\mathbf{w}}' = \mathbf{h}(\mathbf{y}, \mathbf{w}') & \begin{cases} \dot{y}' = x + ay', \\ \dot{z}' = b + z'(x - c). \end{cases}
\end{cases}
\tag{2.27}
$$

In Eqs. (2.27), the *x*-driving is realized because the same driving signal *x* is applied to both subsystems **w** and **w'**. The CS of signals **w** and **w'** is possible if the so-called *conditional Lyapunov exponents* (CLEs) are negative (for more details, see the example in Sec. 4.).

2.4 Active–passive Decomposition

A scheme of *active–passive decomposition*, proposed by Kocarev & Parlitz (1995), can be treated as even a more general drive–response configuration than the *autonomous driver decomposition* described in the previous Sec.. The idea of *active–passive decomposition* consists in rewriting the original autonomous system (2.24) as a non-autonomous one:

$$\dot{\mathbf{z}} = \mathbf{f}(\mathbf{z}, \mathbf{s}(t)),\qquad(2.28)$$

where $\mathbf{s}(t)$ is a driving signal. However, this signal is not an independent external drive, like in Eqs. (2.22) and (2.23), but it is a function of autonomous system variables, i.e., $\mathbf{s}(t) = \mathbf{h}(\mathbf{x})$ or $d\mathbf{s}/dt = \mathbf{h}(\mathbf{x})$. Here, the CS state is also understood as an identical behavior of the non-autonomous system (2.28) and its copy (or copies) representing the response system:

$$\dot{\mathbf{z}}' = \mathbf{f}(\mathbf{z}', \mathbf{s}(t)).\qquad(2.29)$$

In order to exemplify the considered type of the system decomposition, let us substitute the Lorenz system into Eqs. (2.28) and (2.29):

$$\text{drive}\begin{cases}\dot{x} = -\sigma x + s(t),\\ \dot{y} = x(r-z) - y,\\ \dot{z} = xy - bz,\end{cases}$$
$$\text{response}\begin{cases}\dot{x}' = -\sigma x' + s(t),\\ \dot{y}' = x'(r-z') - y',\\ \dot{z}' = x'y' - bz',\end{cases}\qquad(2.30)$$

where the driving signal is $s(t) = \sigma y$. The results presented in Ref. Kocarev & Parlitz (1995) show that for the classical Lorenz oscillator, the CS of the drive with the response system occurs for regular and also for chaotic behavior of the driving signal $s(t)$.

Comparing Eqs. (2.27) and (2.30), we can see that in the Pecora & Carroll scheme the common factor linking the analyzed system with its replica are selected variables (differential equations) of the drive. On the other hand, in the *active–passive decomposition* method both systems are coupled with chosen components of differential equations defining the driving systems. Thus, this latter type of the system decomposition is more flexible in applications than the *autonomous driver* approach because it gives more possibilities to build synchronized dynamical oscillators in numerical simulations and real experiments as well.

Chapter 3

Determination of Complete Synchronization Thresholds

From a viewpoint of practical application, a precise determination of synchronization thresholds belongs to the most fundamental tasks in studies of any synchronization problem. Here, the term synchronization thresholds has been interpreted as the boundaries of the coupling strength ranges where the synchronous state is stable. In such a state, a cooperative motion of coupled oscillators can be observed, i.e., a number of actual degrees of freedom is significantly reduced as compared with a desynchronous case. Then, we can observe the so-called *cluster synchronization* (Kaneko (1994), Terry *et al.* (1999), Pogromsky *et al.* (2002), I. Belykh *et al.* (2003)), i.e., collective dynamics of smaller groups (sub-sets) of oscillators in the network. The extreme case here is a *global synchronization* state when the behavior of the entire network of coupled systems is represented only by the dynamics of its single node. The conditions of such a global asymptotic stability of synchronization are considered in this Chapter.

3.1 Stability of the Synchronous State

In our considerations we have undertaken the following assumptions:

1. All the coupled oscillators are identical, i.e., they are defined by the same set of ODEs with identical values of the system parameters.
2. The synchronous state is located on the invariant manifold. For identical node systems, the manifold invariance results from the constraint of the same vector field components for each block

corresponding to a network node (Pecora *et al.* (2000)). In consequence, a constant row-sum in the connectivity matrix **G** is required for the existence of the synchronization manifold, i.e., $\sum_j G_{ij}$=const. Most of coupling schemes presented in the literature apply the zero row-sum of the vector field (matrix **G**), i.e.,

$$\sum_{j=1}^{N} G_{ij} = 0. \tag{3.1}$$

3.   The *linear coupling* between the network oscillators is applied. For our considerations, it is necessary only in the neighborhood of the synchronization manifold, but in general it can be arbitrary elsewhere (Pecora & Carroll (1998), Pecora *et al.* (2000), Fink *et al.* (2000)).

4.   The same linking function **H** defines the coupling between each pair of network nodes. Remark: an allowance of various **H**-functions in the same network, according to Eqs. (1.1a and 1.1b), requires further development of the network synchronizability theory.

After the above assumptions, Eqs. (1.1a and 1.1b) adopt the forms:

$$\dot{\mathbf{x}} = \mathbf{F}(\mathbf{x}) + \sigma[\mathbf{G} \otimes \mathbf{H}(\mathbf{x})], \tag{3.2a}$$

or

$$\mathbf{x}_{n+1} = \mathbf{F}(\mathbf{x}_n) + \sigma[\mathbf{G} \otimes \mathbf{H}(\mathbf{x}_n)], \tag{3.2b}$$

and the synchronization invariant manifold is defined by $N-1$ constraints:

$$\mathbf{x}_1 = \mathbf{x}_2 = ... = \mathbf{x}_N. \tag{3.3}$$

In order to determine the stability of the synchronous state, various decisive factors can be applied, e.g., Lyapunov exponents or Floquet multipliers. Their negativity is a standard stability criterion, but these tools cannot detect a presence of locally unstable attractors or invariant sets in the phase space (Ashwin *et al.* (1994), Gauthier & Bienfang (1996), Rulkov & Sushchik (1997)). Then, an attractor bubbling or bursting of the system away from the synchronization subspace can occur, especially in the presence of noise or parameter mismatch (Pecora *et al.* (2000)).

3.2 Analytical Approaches for Diagonally Coupled Systems

The first analytical condition for the CS of pairs or regular sets of diagonally coupled identical dynamical systems has been formulated almost simultaneously by Fujisaka & Yamada (1983a), (1983b) and Pikovsky (1984). This condition has been developed both for continuous-time and discrete-time systems with a *symmetrical mutual coupling* (see Sec. 2.1.2). However, the synchronization threshold can be determined analytically also for pairs and networks (regular and random) of non-symmetrically coupled oscillators.

Consider a set of *N* identical dynamical systems with a *diagonal coupling* of an arbitrary configuration between the oscillators. The equations of motion for the system are:

$$\dot{\mathbf{x}}_i = \mathbf{f}(\mathbf{x}_i) + \sum_{r \neq i} \mathbf{D}_{ir}(\mathbf{x}_r - \mathbf{x}_i), \tag{3.4}$$

where $\mathbf{x}_i \in \Re^k$ ($k \geq 3$), $\mathbf{f}(\mathbf{x}_i)$ is a function which governs the dynamics of each individual oscillator and $\mathbf{D}_{ir}=\text{diag}[d_{ir}, d_{ir}, \ldots, d_{ir}] \in \Re^k$ are *diagonal coupling* matrices defining rates of the coupling between each pair of the subsystems in the network ($i, r = 1, 2, \ldots, N, i \neq r$). Thus, this case can be classified as a uniform (identical values of diagonals) CD coupling (Eq. (2.15a)). For $\mathbf{D}_{ir}=\mathbf{0}$ (a lack of the coupling) each of the subsystems given by Eq. (3.4) evolves on the asymptotically stable attractor $\mathcal{A}$. Since these oscillators are identical, it can be assumed that the solutions of the equation $\dot{\mathbf{x}}_i = \mathbf{f}(\mathbf{x}_i)$ starting from different initial points of the same basin of attraction, represent a set of *N* uncorrelated trajectories evolving on the attractor $\mathcal{A}$ (after a period of transient motion). Let us introduce a new variable:

$$\mathbf{x}_{ij} = \mathbf{x}_i - \mathbf{x}_j \tag{3.5}$$

representing the *trajectory separation* between any pair of oscillators ($j=1, 2, \ldots, N, j \neq i$) in the phase space. The complete synchronization of all subsystems requires a fulfillment of the condition:

$$\lim_{t \to \infty} \left\| \mathbf{x}_{ij}(t) \right\| = 0, \quad \forall i, j, \tag{3.6}$$

for each pair of nodes. Putting Eq. (3.4) into the first time derivative of Eq. (3.5) and taking the *diagonal coupling*, we obtain:

$$\dot{\mathbf{x}}_{ij} = \mathbf{f}(\mathbf{x}_i) - \mathbf{f}(\mathbf{x}_j) - (d_{ij} + d_{ji})\mathbf{I}_k \mathbf{x}_{ij}$$
$$- \sum_{r \neq i,j} d_{ir} \mathbf{I}_k \mathbf{x}_{ir} + \sum_{r \neq i,j} d_{jr} \mathbf{I}_k \mathbf{x}_{jr}, \tag{3.7}$$

where $\mathbf{I}_k$ represents the $k{\times}k$ unit matrix. After using the following substitutions:

$$\mathbf{x}_{jr} = \mathbf{x}_{ir} - \mathbf{x}_{ij},$$
$$c_{ij} = -(d_{ij} + d_{ji}) - \sum_{r \neq i,j} d_{jr},$$
$$b_{jr} = d_{jr} - d_{ir},$$

the coupling terms in Eq. (3.7) can be rewritten in the form:

$$- (d_{ij} + d_{ji})\mathbf{I}_k \mathbf{x}_{ij} - \sum_{r \neq i,j} d_{ir} \mathbf{I}_k \mathbf{x}_{ir} + \sum_{r \neq i,j} d_{jr} \mathbf{I}_k \mathbf{x}_{jr}$$
$$= c_{ij}\mathbf{I}_k \mathbf{x}_{ij} + \sum_{r \neq i,j} b_{jr} \mathbf{I}_k \mathbf{x}_{ir}. \tag{3.8}$$

For small *trajectory separation* distances, it can be treated as a perturbation of the synchronous state $\mathbf{x}_i = \mathbf{x}_j$. Hence, in the vicinity of the synchronization subspace (Eq. (3.3)), the following linearized relation can be assumed:

$$\mathbf{f}(\mathbf{x}_i) - \mathbf{f}(\mathbf{x}_j) = \mathbf{J}(t)\mathbf{x}_{ij}. \tag{3.9}$$

The matrix $\mathbf{J}(t)$ in Eq. (3.10) is the Jacobi matrix of any separated node $\dot{\mathbf{x}}_i = \mathbf{f}(\mathbf{x}_i)$ calculated in the base point $\mathbf{x}_i(t)$. Substituting Eqs. (3.8) and (3.9) into Eq. (3.7), we can describe the collection of *trajectory separation* between the given i-th and any other j-th node in the block matrix form:

$$
\begin{pmatrix} \dot{\mathbf{x}}_{i1} \\ \dot{\mathbf{x}}_{i2} \\ \vdots \\ \dot{\mathbf{x}}_{i\,N-1} \end{pmatrix}
=
\left[
\begin{pmatrix}
\mathbf{J}(t) & 0 & \cdots & 0 \\
0 & \mathbf{J}(t) & \cdots & 0 \\
\vdots & \vdots & \ddots & \vdots \\
0 & 0 & \cdots & \mathbf{J}(t)
\end{pmatrix}
\right.
$$
$$
\left.
+
\begin{pmatrix}
c_{i1} & b_{12} & \cdots & b_{1\,N-1} \\
b_{21} & c_{i2} & \cdots & b_{2\,N-1} \\
\vdots & \vdots & \ddots & \vdots \\
b_{N-1\,1} & b_{N-1\,2} & \cdots & c_{i\,N-1}
\end{pmatrix}
\otimes \mathbf{I}_k
\right]
\begin{pmatrix} \mathbf{x}_{i1} \\ \mathbf{x}_{i2} \\ \vdots \\ \mathbf{x}_{i\,N-1} \end{pmatrix}. \tag{3.10}
$$

The above system (Eq. (3.10)) can be block diagonalized. Then, we obtain a *j*-number of separated blocks:

$$\dot{\mathbf{y}}_j = \left(\mathbf{J}(t) - \alpha_j \mathbf{I}_k\right)\mathbf{y}_j, \tag{3.11}$$

where α_j is the *j*-th eigenvalue of the coupling coefficients matrix in Eq. (3.10) and $\mathbf{y}_j$ is $\mathbf{x}_{ij}$ transformed to the eigencoordinates. The solution of Eq. (3.11) initializing from $\mathbf{y}_j(0)$ can be written in the form:

$$\mathbf{y}_j(t) = \left(\exp(-\alpha_j t)\mathbf{Y}(t)\right)\mathbf{y}_j(0), \tag{3.12}$$

where $\mathbf{Y}(t)$ is the fundamental matrix of the equation $\dot{\mathbf{y}}_j = \mathbf{J}(t)\mathbf{y}_j$. From Eq. (3.12), the following solution governing the time evolution of the *trajectory separation* results:

$$\left\|\mathbf{y}_j(t)\right\| = \left\|\mathbf{y}_j(0)\right\| \exp\left[(\lambda_1 - \alpha_j)t\right], \tag{3.13}$$

where λ_1 is the largest Lyapunov exponent (LLE) of the node system. Note that the components connected to the remaining Lyapunov exponents of the node system disappear with the rate of $\exp(-|\lambda_1 - \lambda_m|t)$, where $m=2, 3,....k$. From Eq. (3.13) we see that the norm of the *trajectory separation* tends to zero, i.e., it approaches the synchronization manifold (Eq. (3.3)), if the exponent $\lambda_1 - \alpha_j$ is negative. Hence, the synchronizabilty of the network composed of systems (3.4) is determined with the following theorem:

Theorem 3.1 *If λ_1 is the largest Lyapunov exponent characterizing the time evolution on the attractor $\mathcal{A}$ of the network node system* (3.4) *and α_j represents the eigenvalues of the coupling coefficients matrix* (Eq. (3.10)), *then the fulfillment of the inequality:*

$$\alpha_j > \lambda_1, \quad \forall j, \tag{3.14}$$

allows for the complete synchronization of all the network nodes.

3.2.1 *Two oscillators*

Let us now consider a simple system composed of two diagonally coupled oscillators, according to the scheme presented in Fig. 2.1a. Such a system can be described in the following way:

$$\dot{\mathbf{x}}_1 = \mathbf{f}(\mathbf{x}_1) + d_{12}\mathbf{I}_k(\mathbf{x}_2 - \mathbf{x}_1),$$
$$\dot{\mathbf{x}}_2 = \mathbf{f}(\mathbf{x}_2) + d_{21}\mathbf{I}_k(\mathbf{x}_1 - \mathbf{x}_2),$$

(3.15)

where $\mathbf{x}_1, \mathbf{x}_2 \in \mathfrak{R}^k$. The corresponding variational equation of *trajectory separation* (Eq. 3.11)) takes a form:

$$\dot{\mathbf{z}} = \mathbf{J}(t) - (d_{12} + d_{21})\mathbf{I}_k \mathbf{z},$$

(3.16)

where $\mathbf{z} = \mathbf{x}_1 - \mathbf{x}_2$.

Hence, the eigenvalue $\alpha_{12} = d_{12} + d_{21}$ and the following CS condition of the twin diagonally coupled subsystems (3.15) results from theorem 3.1:

$$d_{12} + d_{21} > \lambda_1.$$

(3.17)

According to Eq. (3.13) one can conclude the time evolution of the norm of the *trajectory separation* can be represented by the equation:

$$\|\mathbf{z}\| = \|\mathbf{z}(0)\| \exp(\lambda_1 t) \exp(-(d_{12} + d_{21})t).$$

(3.18)

The form of Eq. (3.18) perfectly illustrates the mechanism of synchronization of chaotic subsystems (3.15). It can be seen that this process is a product of two independent dynamical effects:

1. exponential divergence of the neighboring trajectories at the rate determined by the positive LLE,
2. exponential convergence of the neighboring trajectories at the rate determined by the sum of the *diagonal coupling* coefficients.

The first of the above effects, which is later called the *effect of the exponential divergence*, is described by a general variational equation:

$$\dot{\mathbf{z}} = \mathbf{J}(t)\mathbf{z}.$$

(3.19)

A geometrical interpretation of the effect is illustrated in Fig. 3.1a, in the form of a 2-dimensional representation of the phase space in the *trajectory separation* coordinate system, represented with the axes z_1 and z_2. The axes $\varepsilon_{1,2}(t)$ represent momentary eigendirections of the ellipsoid $S(t)$. The axis $\varepsilon_1(t)$ in this picture represents the direction of the phase space connected with the positive Lyapunov exponent λ_1, while the time evolution along the axis $\varepsilon_2(t)$ is characterized by the negative exponent λ_2. Furthermore, for the simplification purposes, an agreement of the axes

$z_{1,2}(t)$ and $\varepsilon_{1,2}(t)$ at the time $t=0$ is assumed. Repulsion and attraction of the neighboring trajectories in different, orthogonal directions of the phase space leads, after time t, to a transformation of the initial conditions sphere $S(0)$ (broken line) into the ellipsoid $S(t)$ (continuous line) mentioned in Chapter 1, causing also an increase of the *trajectory separation* $\mathbf{z}(t)$ (continuous line, dependent on λ_1) with respect to the initial vector $\mathbf{z}(0)$ (broken line). Also a change of the space orientation of the main axes of the ellipsoid takes place.

The second of the above-mentioned dynamical effects (*the coupling effect* — hereafter) is defined by the equation:

$$\dot{\mathbf{z}} = -(d_{12} + d_{21})\mathbf{I}_k\mathbf{z} . \tag{3.20}$$

The dynamics of the *coupling effect* is illustrated in Fig. 3.1b. Similar to the *exponential divergence effect*, the volume of the ellipsoid $S(t)$ in the phase space converges to zero in time. But in this case the process is accompanied by diminishing *the trajectory separation vector* at the rate of $\exp(-(d_{12}+d_{21})t)$. This happens due to the fact that the system described in Eq. (3.20) is a simple, linear system with the negative divergence $\text{div}(\mathbf{Dz}) = -k(d_{12}+d_{21})$, and possesses an attractor in the form of a fixed point $\mathbf{z}=\mathbf{0}$. Furthermore, the ellipsoid $S(t)$ preserves its original, spherical shape in the evolution time (nevertheless it is compressed), because the same values of all elements on the main diagonal of the matrix $(d_{12}+d_{21})\mathbf{I}_k$ cause that the scale of an influence of the *coupling effect* is identical in all directions of the phase space.

The isotropic character of an influence of the *coupling effect* in the phase space causes that the synchronization condition (3.17) is fulfilled, even for temporary variable orientation of the ellipsoid $S(t)$ in the phase space, because the orientation does not have any influence on the mutual interaction of the *exponential divergence effect* with the *coupling effect*. It is easy to realize by comparing the mechanisms in both the mentioned dynamical effects, which is illustrated in Figs. 3.1a and 3.1b.

The mutual interaction of these dynamical effects takes place only in the direct vicinity of the synchronization subspace $\mathbf{x}_1=\mathbf{x}_2$, due to the limited influence of the linearized *exponential divergence effect* (the influence of the linear *coupling effect* covers the whole phase space). The fulfillment of inequality (3.17) causes that the subspace becomes a stable attractor. In the opposite case ($\lambda_1 > d_{12} + d_{21}$), the manifold $\mathbf{x}_1 = \mathbf{x}_2$ is a repeller, so the synchronization becomes impossible.

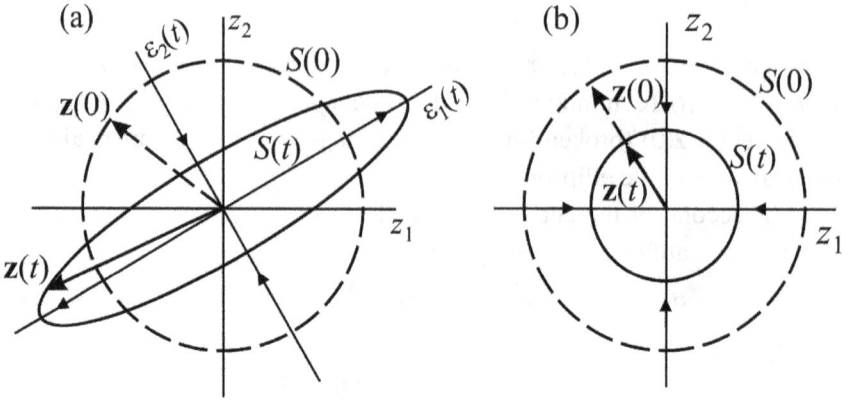

Fig. 3.1 Geometrical interpretation of the *effect of the exponential divergence* (a), and the *diagonal coupling effect* (b).

3.2.2 *Networks of coupled oscillators — diagonal synchronization stability matrix*

The analysis presented above in this Sec. has confirmed that the *diagonal coupling* of identical strange attractors leads to the linear correlation between the LLE and overall coupling coefficient, which can be used for the estimation of the synchronization threshold. An exact determination of the synchronization condition can be done analytically only in some simple cases of coupling configurations, e.g., *symmetrical* or *global coupling*. A more complex structure of the network requires an application of advanced mathematical and numerical techniques (see Secs. 3.3–3.6). However, an important property of the network synchronization with the diagonal type of the coupling results from the investigations carried out. Namely, if only two above-mentioned "parameters of order" play the dominant role in the synchronization process, then we can substitute the node system by any other oscillator characterized by the same value of the LLE without an influence on the process of the network synchronization and the level of the synchronization threshold. This property can be used to simplify the mathematical description of dynamics of chaotic networks. We can reduce the system under consideration (Eq. (3.4)) to the linear, one-dimensional case ($x_i \in \mathfrak{R}^1$), which allows us to determine the synchronization threshold on the basis of linear stability analysis. In order to preserve properties of the original system in the simplified system, two conditions of transformation have to be fulfilled:

1. the substituted system in $\mathfrak{R}^1$ is characterized by the same value of LLE as the original one,
2. original and simplified systems have the identical configuration of coupling.

The presented approach can be applied for continuous-time systems as well as for discrete-time systems (Stefanski *et al.* (2004)).

3.2.2.1 *Continuous-time systems*

In order to construct a linear model of the system (Eq. (3.4)) with one-dimensional nodes, we use the substitutions:

$$\mathbf{f}(\mathbf{x}) = \lambda_1 x, \tag{3.21a}$$

$$\mathbf{D}_{ij} = d_{ij}. \tag{3.21b}$$

Note that the assumed linear system $\dot{x} = \lambda_1 x$ has no attractor if $\lambda_1 > 0$, but it is not a problem here. The most important is the fact that the solution to such a linear system grows exponentially (nearby trajectories diverge with a rate λ_1), which is required by the first condition of the system transformation. Substituting Eqs. (3.21a) and (3.21b) into Eq. (3.4), we obtain a model for the network of one-dimensional systems:

$$\dot{x}_i = \lambda_1 x_i + \sum_{j=1}^{N} d_{ij}(x_j - x_i). \tag{3.22}$$

This simplified model can be rewritten in the vector form:

$$\dot{\mathbf{X}} = \lambda_1 \mathbf{X} + \mathbf{G} \mathbf{X}, \tag{3.23}$$

where $\mathbf{X} = [x_1, x_2, \ldots, x_N]^T$, and

$$G = \begin{bmatrix} -\sum_{j=1}^{N} d_{1j} & d_{12} & \cdots & d_{1N} \\ d_{21} & \ddots & & \vdots \\ \vdots & & \ddots & d_{N-1\,N} \\ d_{N1} & \cdots & d_{N\,N-1} & -\sum_{j=1}^{N} d_{Nj} \end{bmatrix} \qquad (3.24)$$

is the connectivity matrix.

Let us now introduce the *trajectory separation* between an arbitrarily chosen base subsystem and any other j-th oscillator of the network. If we mark the base subsystem by subscript "1", we obtain:

$$x_{1j} = x_1 - x_j,$$
$$x_j - x_r = x_{1r} - x_{1j}. \qquad (3.25)$$

where $j, r = 2, 3, \ldots, N$. Subtracting the remaining subsystems from the base node and applying the introduced substitutions (Eqs. (3.25)), we can rewrite the simplified system in the $(N-1)$-dimensional form where *trajectory separation* variables are given clearly:

$$\dot{Y} = SY, \qquad (3.26)$$

where $Y = [x_{12}, x_{13}, \ldots, x_{1N}]^T \in \Re^{N-1}$ and $(N-1) \times (N-1)$ matrix S assumes the form:

$$S =$$

$$\begin{bmatrix} \lambda_1 - \left(d_{12} + \sum_{j=1}^{N} d_{2j} \right) & \cdots & d_{2k} - d_{1k} & \cdots & d_{2N} - d_{1N} \\ \vdots & \ddots & \vdots & \cdot\cdot\cdot & \vdots \\ d_{i2} - d_{12} & \cdots & \lambda_1 - \left(d_{1k} + \sum_{j=1}^{N} d_{ij} \right) & \cdots & d_{iN} - d_{1N} \\ \vdots & \cdot\cdot\cdot & \vdots & \ddots & \vdots \\ d_{N2} - d_{12} & \cdots & d_{Nk} - d_{1k} & \cdots & \lambda_1 - \left(d_{1N} + \sum_{j=1}^{N} d_{Nj} \right) \end{bmatrix} \qquad (3.27)$$

where the indices i and k enumerate rows and columns, respectively. The system (Eq. (3.27)) now incorporates only transverse dynamics to the synchronization hyper-plane. Therefore, the complete synchronization of all subsystems of the system (Eq. (3.23)) takes place if the critical point of the *trajectory separation* $\mathbf{Y}=\mathbf{0}$ is a stable attractor. Such a situation occurs if real parts of all eigenvalues of the matrix (Eq. (3.27)) are negative. Thus, in agreement with the above assumptions, we can formulate the synchronization condition for general case of the network of chaotic time-continuous systems (Eq. (3.4)) in the following form:

$$\mathrm{Re}(s_i) < 0, \tag{3.28}$$

where s_i ($i=1, 2, \ldots, N-1$) are eigenvalues of the matrix $\mathbf{S}$, which we named the *diagonal synchronization stability matrix* (DSSM) due to its universal character, i.e., the form of DSSM depends only on the network coupling configuration and the LLE of the dynamical system considered as a network node. The DSSM can be constructed directly from the connectivity matrix (Eq. (3.24)), according to the model formula given by (Eq. (3.27)). In general case, we can choose any node of the network as the base to define the DSSM, because it is of no significance for the results of synchronization stability analysis.

3.2.2.2 *Discrete-time systems*

The system analogous to Eq. (3.4) but consisting of N diffusively coupled identical maps is described as follows:

$$\mathbf{x}_{n+1}^i = \mathbf{f}\left(\mathbf{x}_n^i\right) + \sum_{j=1}^{N} d_{ij} \mathbf{I}_k \left[\mathbf{f}\left(\mathbf{x}_n^j\right) - \mathbf{f}\left(\mathbf{x}_n^i\right)\right], \tag{3.29}$$

where $\mathbf{x}_n^i \in \mathfrak{R}^k$ ($k \geq 1$) and $\mathbf{I}_k$ represents the $k \times k$ unit matrix. We obtain the simplified version of Eq. (3.29) by applying the simplest discrete-time system:

$$x_{n+1} = \exp(\lambda_1) x_n, \tag{3.30}$$

which fulfills the first condition of the system simplification, i.e., the LLE of the map given by Eq. (3.30) is equal to λ_1. Using Eqs. (3.21b) and (3.30), the system under consideration (Eq. (3.29)) is reduced to the form analogous to Eq. (3.22) but described by the following set of difference equations:

$$x_{n+1}^i = \exp(\lambda_1)x_n^i + \sum_{j=1}^N d_{ij}\left[\exp(\lambda_1)x_n^j - \exp(\lambda_1)x_n^i\right], \qquad (3.31)$$

or in the vector form:

$$\mathbf{X}_{n+1} = \exp(\lambda_1)(1+\mathbf{G})\mathbf{X}_n. \qquad (3.32)$$

Substituting Eq. (3.25) into Eq. (3.31) and proceeding in the way analogous to the continuous-time system approach shown above, we formulate the difference equations of *trajectory separation* evolution:

$$\mathbf{Y}_{n+1} = \mathbf{M}\mathbf{Y}_n, \qquad (3.33)$$

and a version of the DSSM for maps:

$$\mathbf{M} = \exp(\lambda_1)\times$$

$$\begin{bmatrix} 1-\left(d_{12}+\displaystyle\sum_{j=1}^N d_{2j}\right) & \cdots & d_{2j}-d_{1j} & \cdots & d_{2N}-d_{1N} \\ \vdots & \ddots & \vdots & \ddots & \vdots \\ d_{i2}-d_{12} & \cdots & 1-\left(d_{1j}+\displaystyle\sum_{j=1}^N d_{ij}\right) & \cdots & d_{iN}-d_{1N} \\ \vdots & \ddots & \vdots & \ddots & \vdots \\ d_{N2}-d_{12} & \cdots & d_{Nj}-d_{1j} & \cdots & 1-\left(d_{1N}+\displaystyle\sum_{j=1}^N d_{Nj}\right) \end{bmatrix}.(3.34)$$

Hence, the synchronization threshold for the ensembles of chaotic maps with a regular or random configuration of the coupling is defined by the inequality:

$$|\mu_i| < 1, \quad \forall i, \qquad (3.35)$$

where μ_i ($i=1, 2, \ldots, N-1$) are eigenvalues of the DSSM (Eq. (3.34)).

3.3 Arbitrary Linking Function

The CD coupling considered in the previous Sec. 3.2 is very convenient for analytical considerations because it allows us to "linearize" any synchronization problem. However, in many physical systems, such a way of connection cannot be realized, e.g., in mechanical systems. Thus, there appeared a need to elaborate more advanced tools for stability analysis of the synchronous state including any possible linking functions **H,** like the PD, ND and other combined cases. In recent years many researchers (mentioned and cited in previous Chapters) have undertaken attempts to solve this synchronization threshold problem. Many of the proposed approaches have been applied for describing the synchronization problem for particular coupling configurations as well as for more general cases. Most of the existing works on the network synchronization refer to a regular, symmetrical structure of the coupling. However, a nonsymmetrical (in particular unidirectional) and random coupling configuration has been also considered in some papers (Heagy *et al.* (1994), Wu & Chua (1994), (1995a), (1995b), Gade *et al.* (1995), Gade (1996), Brown & Rulkov (1997), Dmitriev *et al.* (1997), Pecora & Carroll (1998), Watts & Strogatz (1998), Pecora *et al.* (2000), Belykh *et al.* (2004), Stefanski *et al.* (2004), Duan *et al.* (2008)). In most of these works the properties of the eigenvalue spectrum of the connectivity matrix have been exploited for quantifying the synchronizability of coupled systems. Especially noteworthy here is the concept called *Master Stability Function* (MSF) introduced by Pecora & Carroll (1998), which allows one to solve the network synchronization problem for any set of coupling weights and connections, and any number of coupled oscillators. Similar techniques have been used in other works mentioned above. There exist also some proposals combining the MSF with other theories, e.g., Gershgörin disk theory (Chen Y. *et al.* (2003)). Equally interesting solutions are applications of the graph theory to configurations of oscillators, e.g., an approach developed by Wu & Chua (1994), (1995a), (1995b), complementary graphs (Duan *et al.* (2008)) or a connection graph based stability method introduced by Belykh *et al.* (2004). This last concept makes it possible to estimate the synchronization threshold without calculation of Lyapunov exponents and eigenvalues of the connectivity matrix. On the other hand, in the case of decomposed (Sec. 2.3 and 2.4) or externally driven systems (Sec. 2.2), their synchronizability can be quantified with the so-called *conditional* or *response Lyapunov exponents* (Pecora & Carroll (1991), Stefanski & Kapitaniak (2003a), Stefanski (2008)). Several examples of the above-mentioned methods and tools have been described below in this Sec.

3.3.1 *Master stability function*

The concept of the MSF can be treated as the representative one among the stability criterions based on the eigenvalue spectrum of the connectivity matrix. This idea is presented here also because it has been applied as a tool for the investigation of synchronization thresholds described in Chapter 4.

The MSF can be determined in three different cases:

1. numerical calculations of the stability criterion for time-continuous systems,
2. numerical calculations of the stability criterion for maps,
3. direct numerical or experimental investigations of synchronous ranges in a two or three oscillators probe.

3.3.1.1 *The MSF for continuous-time systems*

In order to explain the concept of the MSF for time-continuous systems, let us take under consideration system (3.2a) with the assumptions listed in Sec. 3.1. Then, any set of coupled oscillators (see Fig. 2.2d) can be described in the more detailed block form:

$$
\begin{pmatrix} \dot{\mathbf{x}}_1 \\ \dot{\mathbf{x}}_3 \\ \vdots \\ \dot{\mathbf{x}}_N \end{pmatrix} = \begin{pmatrix} \mathbf{f}(\mathbf{x}_1) \\ \mathbf{f}(\mathbf{x}_2) \\ \vdots \\ \mathbf{f}(\mathbf{x}_N) \end{pmatrix} + \sigma \left[\begin{pmatrix} G_{11} & G_{12} & \cdots & G_{1N} \\ G_{21} & G_{22} & \cdots & G_{2N} \\ \vdots & \vdots & \ddots & \vdots \\ G_{N1} & G_{11} & \cdots & G_{NN} \end{pmatrix} \otimes \mathbf{H} \right] \times \begin{pmatrix} \mathbf{x}_1 \\ \mathbf{x}_2 \\ \vdots \\ \mathbf{x}_N \end{pmatrix} . \quad (3.36)
$$

where the *linear coupling* $\mathbf{H}(\mathbf{x}_i) = \mathbf{H}\mathbf{x}_i$ is applied according to requirement (3) in Sec. 3.1 and $\mathbf{x}_i \in \Re^m$.

As a tool for testing the stability of synchronous state, we have applied Lyapunov exponents. These quantities determine the divergence of nearby trajectories in directions transverse to the synchronization manifold (Eq. (3.3)), so they are called *transversal Lyapunov exponents* (TLEs — λ_T, Anishchenko *et al.* (2001)). Therefore, it requires separating the transverse modes from the longitudinal one in the variational equation. Deriving system (3.36), we obtain:

$$
\begin{pmatrix} \dot{\xi}_1 \\ \dot{\xi}_3 \\ \vdots \\ \dot{\xi}_N \end{pmatrix} = \left[\begin{pmatrix} Df & 0 & \cdots & 0 \\ 0 & Df & \cdots & 0 \\ \vdots & \vdots & \ddots & \vdots \\ 0 & 0 & \cdots & Df \end{pmatrix} + \right.
$$

$$
\left. + \sigma \begin{pmatrix} G_{11} & G_{12} & \cdots & G_{1N} \\ G_{21} & G_{22} & \cdots & G_{2N} \\ \vdots & \vdots & \ddots & \vdots \\ G_{N1} & G_{11} & \cdots & G_{NN} \end{pmatrix} \otimes DH \right] \times \begin{pmatrix} \xi_1 \\ \xi_2 \\ \vdots \\ \xi_N \end{pmatrix}
$$

(3.37)

where ξ_i represents an m-dimensional perturbation of the i-th node, Df is the Jacobi matrix of any node, i.e., the derivative with respect to the first argument of the function $f(x_i)$, the same for all oscillators on the synchronization manifold, and DH is the Jacobian of the linking function H.

The next stage is a diagonalization of Eq. (3.37). Such block diagonalization leads to the uncoupling of variational Eq. (3.37) into blocks like in a mode analysis. After such a block diagonalization of the variational equation, there appear N separated blocks:

$$
\dot{\xi}_k = [Df + \sigma \gamma_k DH)] \xi_k ,
$$

(3.38)

where ξ_k represents different transverse modes of a perturbation from the synchronous state and γ_k represents a k-th eigenvalue of the connectivity matrix G, $k=0, 1, 2, \ldots , N-1$. An orientation of the set of coordinates in the phase space of system (3.36) before and after the diagonalization is depicted in Fig. 3.2. For $k=0$ we have $\gamma_0 = 0$ and Eq. (3.38) is reduced to the variational equation of the separated node system:

$$
\dot{\xi}_k = Df \xi_k ,
$$

(3.39)

corresponding to the longitudinal direction located within the synchronization manifold (the coordinate x_1^G in Fig. 3.2). All other k-th eigenvalues correspond to transverse eigenvectors (the coordinate x_2^G in Fig. 3.2).

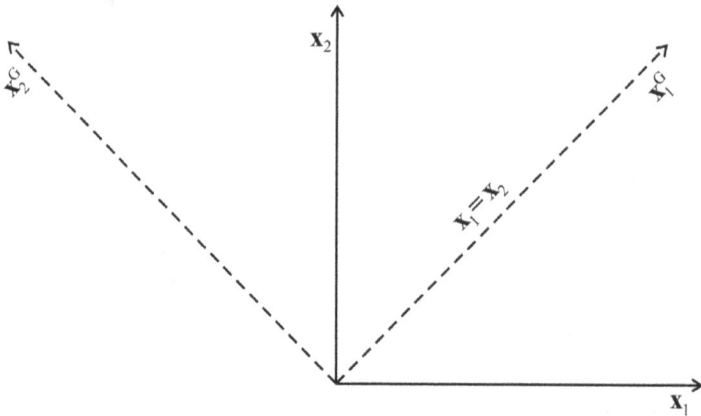

Fig. 3.2. Orientation of the space coordinate systems of the phase space of system (3.36) before (continuous line) and after (broken line) the diagonalization.

In accordance with the MSF concept, a tendency to synchronization of the network is a function of the eigenvalues γ_k. Substituting $\sigma\gamma = \alpha + i\beta$, where $\alpha = \sigma\mathrm{Re}(\gamma)$, $\beta = \sigma\mathrm{Im}(\gamma)$ and γ represents an arbitrary value of γ_k, we obtain the generic variational equation:

$$\dot{\xi} = [D\mathbf{f} + (\alpha + i\beta)D\mathbf{H})]\xi, \tag{3.40}$$

where ξ symbolizes an arbitrary transverse mode. The connectivity matrix $\mathbf{G}$ satisfies a zero row-sum (Eq. (3.1)), so that the synchronization manifold $\mathbf{x}_1 = \mathbf{x}_2 = \ldots = \mathbf{x}_N$ is invariant and all the real parts of eigenvalues γ_k associated with transversal modes are negative ($\mathrm{Re}(\gamma_{k\neq0}) < 0$). Hence, we obtain the following spectrum of the eigenvalues of $\mathbf{G}$: $\gamma_0 = 0 \geq \gamma_1 \geq \gamma_{N-1}$. Now, we can define the MSF as a surface representing the largest TLE λ_T, calculated for generic variational equation (Eq. (3.40)), over the complex numbers plane (α, β). Obviously, the calculation of the MSF requires a simultaneous integration of the node system $d\mathbf{x}_i/dt = \mathbf{f}(\mathbf{x}_i)$. If an interaction between each pair of nodes is mutual and symmetrical, then a real coupling of oscillators takes place (Sec. 2.1.4.1), i.e., $\beta_k = 0$. In such a case, the MSF is reduced to a form of the curve representing the largest TLE as a function of the real number α (see Fig. 3.3) fulfilling the equation:

$$\alpha = \sigma\gamma. \tag{3.41}$$

If all the eigenmodes corresponding to the discrete spectrum of eigenvalues $\sigma\gamma_k$ can be found in the ranges of negative TLE (Fig. 3.3a), then the synchronous state is stable for the considered configuration of couplings. On the other hand, if even only one of the eigenvalues is located in the area of positive TLE (Fig. 3.3b), then the *global synchronization* of all network nodes is unstable but, e.g., an appearence of the *cluster synchronization* is possibile.

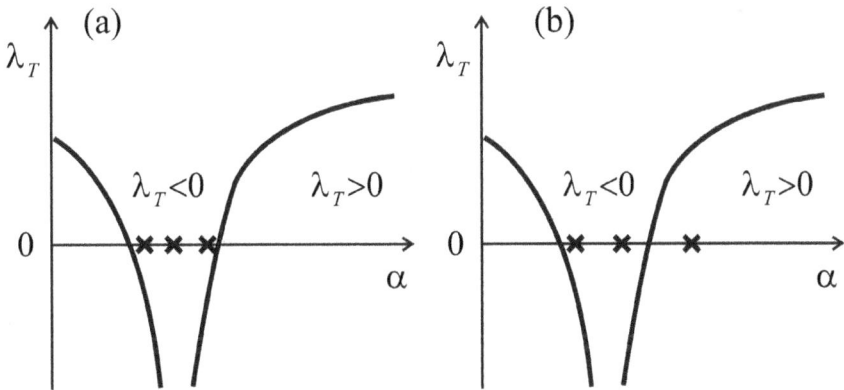

Fig. 3.3. Visualization of any discrete spectrum of real eigenvalues of connectivity matrix on the background of the exemplary MSF plot λ_T (α) representing the synchronous (a) and desynchronous (b) tendency of the network oscillators.

3.3.1.2 *The MSF for maps*

The previous analysis of the MSF can be generalized for the case of coupled maps. The dynamics of the individual node is described by:

$$\mathbf{x}_{n+1}^i = \mathbf{f}(\mathbf{x}_n^i). \tag{3.42}$$

In general, several ways to couple discrete-time systems are possible (Fink *et al.* (2000)), i.e.,

1.   through added arguments,

$$\mathbf{x}_{n+1}^i = \mathbf{f}\left(\mathbf{x}_n^i + \sum_j \sigma G_{ij} \mathbf{H}(\mathbf{x}_n^j)\right), \tag{3.43}$$

e.g., the *dissipative coupling* of maps (Eq. (2.20)),

2.   through added functions

$$\mathbf{x}_{n+1}^i = \mathbf{f}(\mathbf{x}_n^i) + \sum_j \sigma G_{ij} \mathbf{H}(\mathbf{x}_n^j), \tag{3.44}$$

the general functional coupling,

$$\mathbf{x}_{n+1}^i = \mathbf{f}\left(\mathbf{x}_n^i, \{\mathbf{H}(\mathbf{x}_n^j)\}\right), \tag{3.45}$$

where $i, j = 1, 2, \dots, N$.

The form of variational equations are similar for the cases presented above. Therefore, for appropriate assumptions all three forms are in fact the same when the variational equations are evaluated on the synchronization manifold. Thus, let us consider then only the first coupling scheme (Eq. (3.43)) as a representative example. Let $\xi^i \in \Re^m$ be a variation of the i-th node around the completely synchronous solution $\mathbf{x}_n^1 = \dots = \mathbf{x}_n^N = \mathbf{s}_n$. Then, the variational equation of system (3.43) is:

$$\xi_{n+1}^i = [D\mathbf{f}(\mathbf{s}_n) + \sum_j \sigma G_{ij} D\mathbf{f}(\mathbf{s}_n) D\mathbf{H}(\mathbf{s}_n)]\xi_n^i, \tag{3.46}$$

where $D\mathbf{f}(\mathbf{s}_n)$ and $D\mathbf{H}(\mathbf{s}_n)$ are Jacobians of the node map and the output function, respectively, evaluated on the synchronization manifold. Hence, for the collection of variations $\xi_n = (\xi_n^1, \xi_n^2, \dots, \xi_n^N)$, we have:

$$\xi_{n+1} = [\mathbf{1}_N \otimes D\mathbf{f}(\mathbf{s}_n) + \sigma \mathbf{G} \otimes (D\mathbf{f}(\mathbf{s}_n) D\mathbf{H}(\mathbf{s}_n))]\xi_n. \tag{3.47}$$

The connectivity matrix satisfies

$$\Sigma_{j=1}^N G_{ij} = 0,$$

like in the case of continuous-time systems. System (3.47) can be block diagonalized with the blocks:

$$\eta_{n+1}^k = D\mathbf{f}(\mathbf{s}_n)[1 + \sigma \gamma_k D\mathbf{H}(\mathbf{s}_n)]\eta_n^k, \tag{3.48}$$

where $\eta^k \in \Re^m$ are new variations after diagonalization and γ_k are eigenvalues of the coupling matrix $\mathbf{G}$. Now, the generic variational equation (an equivalent of Eq. (3.40)) can be defined as follows:

$$\eta_{n+1} = D\mathbf{f}(\mathbf{s}_n)[1 + (\alpha + i\beta)D\mathbf{H}(\mathbf{s}_n)]\eta_n, \tag{3.49}$$

where η_n represents any transverse mode η_n^k. Finally, the map version of the MSF can be calculated in the way analogous to the differential equation case, on the basis of Eqs. (3.49) and (3.42).

3.3.1.3 *Three or two oscillators probe*

The calculation of the MSF according to the above-described procedures is not a very complicated task when differential or difference equations of the node system are known and differentiable and the stability criterion can be strictly defined. On the other hand, if the numerical model of the node oscillator is inaccurate, discontinuous, or even unknown, then the numerical calculation of the MSF can be very difficult or even impossible. In such cases the MSF can be evaluated not straightforward, i.e., without use of variational Eqs. (3.40) and (3.49). This technique, called *three oscillators universal probe*, have been introduced by Fink *et al.* (2000). They have designed a simple three-oscillator array where *real symmetric* (Eqs. (2.7) for $N=3$) and *imaginary anti-symmetric* (Eqs. (2.13) and (2.14)) kinds of the coupling are realized simultaneously (see the scheme shown in Fig. 3.4). This is the simplest configuration of three nodes ($i = 1, 2, 3$):

$$\dot{\mathbf{x}}_i = \mathbf{f}(\mathbf{x}_i) + \sigma\left(\frac{\mu}{3}[\mathbf{H}(\mathbf{x}_{i+1}) + \mathbf{H}(\mathbf{x}_{i-1}) - 2\mathbf{H}(\mathbf{x}_i)]\right.$$
$$\left. + \frac{\rho}{\sqrt{3}}[\mathbf{H}(\mathbf{x}_{i+1}) - \mathbf{H}(\mathbf{x}_{i-1})]\right), \tag{3.50}$$

which allows us to estimate the MSF over the entire complex plane via the direct detection of the synchronization process in real or numerical experiments. The parameters μ and ρ are *real* and *imaginary coupling* factors, respectively, and the denominators 3 and $\sqrt{3}$ have been used in order to simplify the further notation. The variational equation of the system (3.50) is:

$$\begin{pmatrix} \dot{\xi_1} \\ \dot{\xi_2} \\ \dot{\xi_3} \end{pmatrix} = \left[\mathbf{1_3} \otimes D\mathbf{f} + \sigma \begin{pmatrix} \dfrac{-2\mu}{3} & \dfrac{\mu}{3}+\dfrac{\rho}{\sqrt{3}} & \dfrac{\mu}{3}-\dfrac{\rho}{\sqrt{3}} \\ \dfrac{\mu}{3}-\dfrac{\rho}{\sqrt{3}} & \dfrac{-2\mu}{3} & \dfrac{\mu}{3}+\dfrac{\rho}{\sqrt{3}} \\ \dfrac{\mu}{3}+\dfrac{\rho}{\sqrt{3}} & \dfrac{\mu}{3}-\dfrac{\rho}{\sqrt{3}} & \dfrac{-2\mu}{3} \end{pmatrix} \otimes D\mathbf{H} \right] \begin{pmatrix} \xi_1 \\ \xi_2 \\ \xi_3 \end{pmatrix}. \quad (3.51)$$

Owing to appropriately chosen denominators in the coupling terms of Eq. (3.50), we have the following eigenvalues of the connectivity matrix: $\gamma_0 = 0$, $\gamma_{1,2} = -\mu \pm i\rho$. Thus, the generic variational equation for determining the MSF of the probe system is as follows:

$$\dot{\xi} = [D\mathbf{f} + \sigma(\mu \pm i\rho)D\mathbf{H})]\xi, \quad (3.52)$$

where $\sigma, \mu, \rho \in \Re_+$. Comparing Eq. (3.40) with the above Eq. (3.52), we can see the relations $\alpha = \sigma\mu$ and $\beta = \sigma\rho$.

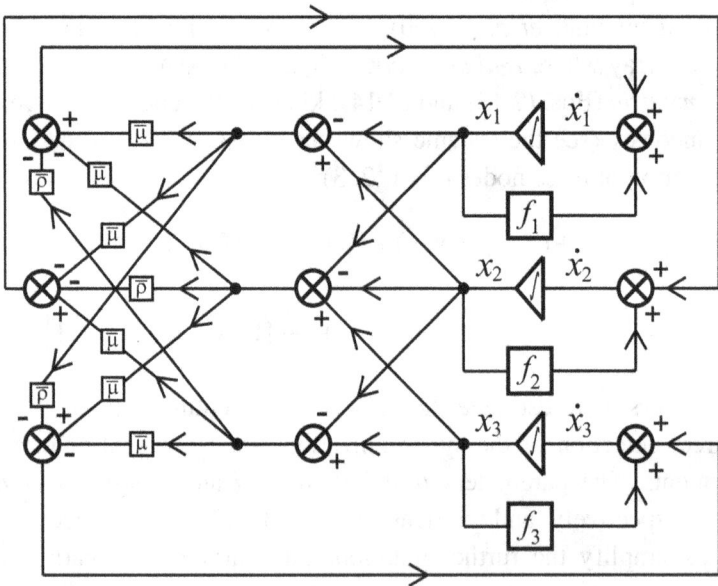

Fig. 3.4. General scheme of connections in *three oscillators universal probe* (see Eq. (3.51)); $\overline{\rho} = \rho/\sqrt{3}$ and $\overline{\mu} = \mu/3$.

Consequently, in order to evaluate the MSF using the technique under consideration, we have to couple three identical oscillators according to the connectivity scheme from Eq. (3.51). The next step is the investigation of the CS of this triple system when the real μ and imaginary ρ factors vary in some boundaries circumscribing a complex number plane. The area on this plane where the CS (or the ICS in the case of real experiment) can be observed is an equivalent of the stable synchronous region of the MSF, where the largest TLE is negative. Thus, the main advantage of the *three oscillators universal probe* is the possibility of the experimental realization and its usefulness for discontinuous and non-differentiable systems. Moreover, sometimes this technique can have a significant practical advantage over the algorithmic version of the MSF even in numerical investigations of continuous oscillators because the numerical detection of synchronization can be faster in realization than the calculation of Lyapunov or Floquet exponents.

It is also possible to execute a *three oscillators probe* in a simpler version with a reduced number of connections between the oscillators (Wu (2002)) However, the easiest application of this method is *two oscillators probe*, which is enough when only a *real coupling* is introduced to connect the oscillators in the network and the connectivity matrix is symmetrical. Such a situation takes place, e.g., in systems of coupled mechanical oscillators. Then, Eq. (3.51) is reduced to:

$$
\begin{pmatrix} \dot{\xi}_1 \\ \dot{\xi}_2 \end{pmatrix} = \left[\mathbf{1}_2 \otimes Df + \sigma \begin{pmatrix} -\dfrac{\mu}{2} & \dfrac{\mu}{2} \\ \dfrac{\mu}{2} & -\dfrac{\mu}{2} \end{pmatrix} \otimes DH \right] \begin{pmatrix} \xi_1 \\ \xi_2 \end{pmatrix}, \qquad (3.53)
$$

where real eigenvalues of **G** are $\gamma_0 = 0$, $\gamma_1 = -\mu$. Consequently, the generic variational equation is now as follows:

$$
\dot{\xi} = (Df - \sigma\mu DH)\xi . \qquad (3.54)
$$

Thus, for $\mu = 1$ we have $\alpha = \sigma$.

3.3.2 *Graph methods*

Another instrument, which is useful in the mathematical analysis of networks, is the algebraic graph theory (Biggs (1993), Wu & Chua (1995a), Belykh *et al.* (2004), Wu (2005), Duan *et al.* (2007), (2008)) because the structure of any network can be represented by the corresponding graph. Recently, there have appeared some examples of application of the graph theory to study the network synchronization problem. Wu & Chua have introduced the idea of *directed interaction graphs* to make the visualization of connections between the nodes easier (Wu & Chua (1995a), (1995b)). The definition of such a graph is following (cited by Wu & Chua (1995a)):

Definition 3.1 *The directed matrix graph* Γ_A *of the* $m \times m$ *matrix* **A** *is associated with the matrix* **A** *as follows: there are* m *vertices in* Γ_A *with an edge from the vertex* j *to the vertex* i *if and only if* A_{ij} *is nonzero.*

A scheme of this kind of graph for the chain of diffusively coupled cells, corresponding to the connectivity matrix given by Eq. (2.10), is shown in Fig. 3.5.

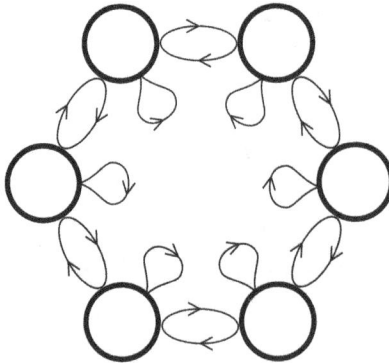

Fig. 3.5. Examlpe of directed matrix graph for diffusively (nearest neighbor) bidirectionally coupled oscillators (Eq. (2.10)).

In turn, other researchers (Atay & Bıyıkoğlu (2005), Duan *et al.* (2008)) have applied the so-called sub-graphs or complementary graphs to study the network synchronizability. An arbitrary graph and its complementary associate are shown in Figs. 3.6a, b. Usually a new edge in the network changes its synchronizability. In general, there exist redundant edges in the network, which not only make no contribution to synchronization but even may reduce the synchronizability. However, it has been shown (Atay & Bıyıkoğlu (2005), Duan *et al.* (2008)) that such additional connections do not decrease the synchronization tendency in networks with disconnected complementary graphs (Figs. 3.6a, b).

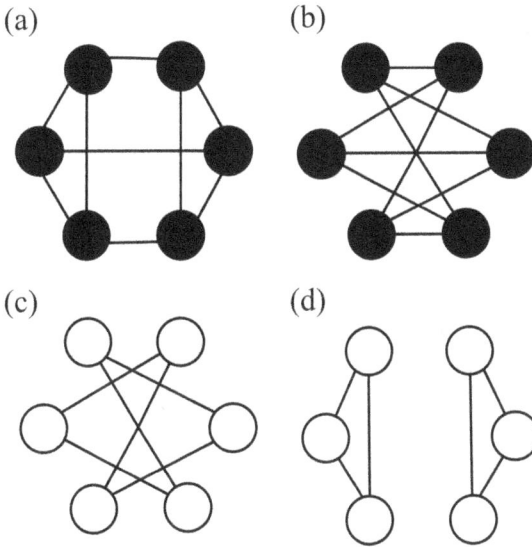

Fig. 3.6. Examples of graphs representing networks connections — (a) G_1 and (b) G_2, and their complementary graphs — (c) G^c_1 and (d) G^c_2, respectively.

A common feature of the above-mentioned graph-theoretic approaches is that they require eigenvalues of the connectivity matrix **G** to be calculated. Thus, in fact these techniques are qualitatively similar to the MSF method. In general, such methods based on **G**-eigenvalues can be used for networks defined by the connectivity matrix of constant components. But in the case of a time-varying coupling, when the matrix **G** is time-dependent, the use of these methods is very difficult and often even impossible.

In recent times, an alternative graph-based method for determining the synchronization threshold in networks of mutually coupled oscillators has been proposed (Belykh *et al.* (2004)), i.e., the so-called *connection graph stability* method (CGS). This technique coalesces the graph theoretical representation of the network with the idea of Lyapunov function. Such a direct link of the synchronization with the graph theory makes it possible to avoid the calculation of **G**-eigenvalues. A significant advantage of this approach is its applicability to time-dependent networks.

In order to approach briefly this method, let us consider again the network of mutually coupled cells (e.g., Eq. (2.8)) in a slightly different notation:

$$\dot{\mathbf{x}}_i = \mathbf{f}(\mathbf{x}_i) + \sum_{j=1}^{N} G_{ij}(t) \mathbf{H} \mathbf{x}_j , \qquad (3.55)$$

where $i = 0, 1, 2, . , N$, $G_{ij}(t)$ represents potentially time-varying components of the connectivity matrix **G**. Here, this matrix defines a graph connecting N nodes by means of l edges. The graph has an edge between two nodes i and j if $G_{ij} = G_{ji} > 0$. The crucial operation characterizing the CGS is to select a path $\mathbf{H}_{ij}$ from the vertex i to the vertex j for any pair nodes (i, j). Next, the total length (sum) of all selected paths passing through the edge k on the connection graph has to be calculated ($k = 0, 1, 2, . , l$). The global CS of network oscillators is guaranteed if the coupling strength is proportional to the maximum value of this sum when k varies. Hence, in accordance with the CGS concept, the sufficient condition for the network *global synchronization* can be defined by the following theorem (cited by Belykh *et al.* (2004)):

Theorem 3.2 *The synchronization manifold of system (3.55) is globally asymptotically stable if:*

$$G_k(t) > \frac{a}{N} b_k(N, l) \qquad (3.56)$$

for k = 0, 1, 2, . , l and for all t. Here, $a = 2G_2^*$ is the double critical coupling strength sufficient for the global synchronization of two oscillators, where G_2^* is the minimum coupling strength between two oscillators such that $\left\| \mathbf{Hx}_2 - \mathbf{Hx}_1 \right\|^2$ is a global Lyapunov function. The quantity $b_k(N,l) = \sum_{j>i; k \in \mathbf{H}_{i,j}}^{N} \left| \mathbf{H}_{i,j} \right|$ is the sum of the lengths of all chosen paths $\mathbf{H}_{ij}$, which pass through a given k that belongs to the coupling graph.

The first stage of the method implementation is the calculation of the coefficient a, i.e., the synchronization threshold for two oscillators coupled according to the linking function $\mathbf{H}$. Next, in order to calculate the value $b_k(N, l)$, we have to select a set of paths { $\left| \mathbf{H}_{ij} \right|$ $i, j = 0, 1, 2, \dots$, N, $j > i$}, for each pair of nodes, determine their lengths $\left| \mathbf{H}_{ij} \right|$ and the number of edges in each $\mathbf{H}_{ij}$. Then, for each edge k of the connection graph, the sum $b_k(N, l)$ can be calculated.

Depending on the choice of paths, we get the lower limit (3.56) for each coupling strength G_k. If all these limits on G_k^* are satisfied, then Theorem 3.2 ensures the CS of all network nodes.

3.3.3 Conditional and response Lyapunov exponents

In the previous Sec. the idea of TLEs has been mentioned as a criterion for stability of the synchronous state. This tool is especially useful in the case of the direct, e.g., *diffusive coupling* between the oscillators via connecting components. Such a connection often causes the transversal (with respect to the synchronization manifold) convergence of trajectories in the phase space. However, in the case of a master–slave connection in decomposed or externally driven systems, the mechanism of synchronization is slightly different. Namely, the synchronization of response oscillators is possible if they "forget their initial conditions" (Boccaletti *et al.* (2002)). Such a situation takes place when the Lyapunov exponents characterizing response subsystems are negative. As it was mentioned above, these exponents have been called *conditional Lyapunov exponents* (CLEs) or *response Lyapunov exponents* or (RLEs).

3.3.3.1 Decomposed systems

In order to approach briefly the concept of CLEs, let us come back to the example of the decomposed Rössler system from Sec. 2.3 (Eq. (2.27)). As the drive in system (2.27), the x variable is applied. Hence, the CLEs are Lyapunov exponents of the uncoupled sub-block yz. They can be calculated from the Jacobian of the yz sub-block, i.e.,

$$\begin{pmatrix} a & 0 \\ 0 & x-c \end{pmatrix}. \tag{3.57}$$

In the classical Rössler system, the parameter a is positive and x is usually much smaller than c. Thus, the maximum CLE of the sub-block (3.57) is also positive because it is equal to a. Therefore, the subsystems of x-driven Rössler oscillators cannot synchronize.

Consider now the y-driven Rössler systems:

$$\dot{x} = -y - z,$$
$$\dot{y} = x + ay,$$
$$\dot{z} = b + z(x - c), \tag{3.58}$$
$$\dot{x}' = -y - z',$$
$$\dot{z}' = b + z'(x' - c),$$

where the uncoupled linearized sub-block xz is as follows:

$$\begin{pmatrix} 0 & -1 \\ z & x-c \end{pmatrix}. \tag{3.59}$$

The eigenvalues of Jacobian (3.59) approximating the CLEs are:

$$\lambda_{1,2} = \frac{1}{2}\left(x - c \pm \sqrt{(c-x)^2 - 4z} \right). \tag{3.60}$$

The variable z in the system under consideration is almost always non-negative. Hence, for x much smaller than c, the real part of the eigenvalues (Eq. (3.60)) is negative. Consequently, the CLEs of y-driven subsystems of the Rössler oscillator are negative, so their synchronization is possible.

3.3.3.2 *Externally driven oscillators*

The general description of the systems with a common external drive is given in Sec. 2.2 in the block matrix form (Eqs. (2.22) and (2.23)). Here, the synchronization mechanism in these systems is explained. This mechanism is described only in a version for continuous-time systems (Eq. (2.22)) due to its complete analogy to the case of maps (Eq. (2.23)).

In order to investigate the synchronizability of the array of externally excited oscillators, the properties of the GS have been employed (Rulkov *et al.* (1995), Abarbanel *et al.* (1996), Kocarev, Parlitz (1996)). We have assumed that all response oscillators are identical. The dynamics of each individual driven oscillator ($\mathbf{x} = \mathbf{x}_i$, $i=1, 2,..., N$) is expressed by the following equations:

$$\dot{\mathbf{e}} = \mathbf{g}(\mathbf{e}) . \tag{3.61a}$$

$$\dot{\mathbf{x}} = \mathbf{f}(\mathbf{x}) + q\mathbf{h}(\mathbf{e}) . \tag{3.61b}$$

Two cases of the driving character can be distinguished, namely:

1. *deterministic driving*, when the driving system is known, i.e., the function $\mathbf{g}$ (Eq. (3.61a)) and the dimension k of the phase space $\mathbf{D}$ of the driving system (*driving subspace*) are strictly defined.
2. *non-deterministic driving*, when dynamics of excitation is governed by a stochastic function $\mathbf{g}$ or even $\mathbf{g}$ and k are completely indefinite and only an output signal from the driving system is known.

In the first case, the solution to systems (3.61a) and (3.61b) exists in the $(k+m)$-dimensional phase space $\mathbf{D} \oplus \mathbf{R}$, where $\mathbf{R}$ is the phase space (*response subspace*) of the response system (Eq. (3.61b)). The solution is characterized by a spectrum of a k-number of *driving Lyapunov exponents* (DLEs) and an m-number of the RLEs. In the second case, determination of Lyapunov exponents is often impossible due to the stochastic or unknown driving.

However, in both cases the solution to the response system (Eq. (3.61b)) can be assumed in the following form:

$$\mathbf{x}(t) = \mathbf{\Phi}[\mathbf{x}(t), \mathbf{x}_0] + \mathbf{\Psi}[\mathbf{e}(t)] , \tag{3.62}$$

where $\boldsymbol{\Phi}$ and $\boldsymbol{\psi}$ represent the functional parts of the solution, which are dependent on and independent of the response sub-system, respectively. In order to examine the synchronization tendency of the response oscillators, let us consider two of them, arbitrarily chosen from system (3.61a), i.e., $\mathbf{x}_i$ and $\mathbf{x}_{i+1}$. The time evolution of the *trajectory separation* between them (*synchronization error*) is described by the equation:

$$\dot{\mathbf{x}}_i - \dot{\mathbf{x}}_{i+1} = \mathbf{f}(\mathbf{x}_i) - \mathbf{f}(\mathbf{x}_{i+1}), \tag{3.63}$$

resulting from Eq. (3.61b). The linearization of Eq. (3.63) leads to the following variational equation:

$$\dot{\xi} = D\mathbf{f}[\mathbf{x}(t), \mathbf{x}_0]\xi, \tag{3.64}$$

where $D\mathbf{f}[\mathbf{x}(t), \mathbf{x}_0]$ is the Jacobi matrix of the response system. On the basis of Eq. (3.64), the RLEs of systems (3.61a) and (3.61b) can be calculated. From Eqs. (3.63) and (3.64) it results that the *synchronization error* tends to zero and the synchronous state is stable if all the RLEs are negative:

$$\lambda_j^R < 0, \tag{3.65}$$

where $j=1, 2,\ldots, m$. Then, the component of solution (3.62) associated with the response of the system $\boldsymbol{\Phi}[\mathbf{x}(t), \mathbf{x}_0]$ tends to zero and there appears a functional relation between the drive and the response systems analogous to Eq. (1.6). Thus, the GS of systems (3.61a) and (3.61b) takes place because the external drive results in the response, forgetting its initial condition.

A two-dimensional visualization of the system phase space is presented in Fig. 3.7. The system given by Eqs. (2.22) or (2.23) can be considered as a set of identical subsystems with a common drive (Stefanski & Kapitaniak (2003), Stefanski (2008)). Hence, it can be assumed that the solutions of Eq. (2.22), starting from different initial conditions, represent independent trajectories $\mathbf{x}_i(t)$ evolving on the same attractor (after a period of the transient motion). Common excitation causes that the *trajectory separation* between these trajectories in the *driving subspace* $\mathbf{D}$ is zero at each moment. If all the RLEs are negative, then there exists a point in the *response subspace* which is a stable *sub-attractor*. This point is a trace of the system trajectory in the *response subspace* **R**. It causes that trajectories starting from different points of the basin of attraction evolve to the same state and oscillators will synchronize (Fig. 3.7a). In other words, an invariant subspace representing the synchronized state $x_1 = x_2 = \ldots = x_N$ is a stable attractor. Such a synchronization is caused by the common drive only and it occurs without any additional coupling between oscillators.

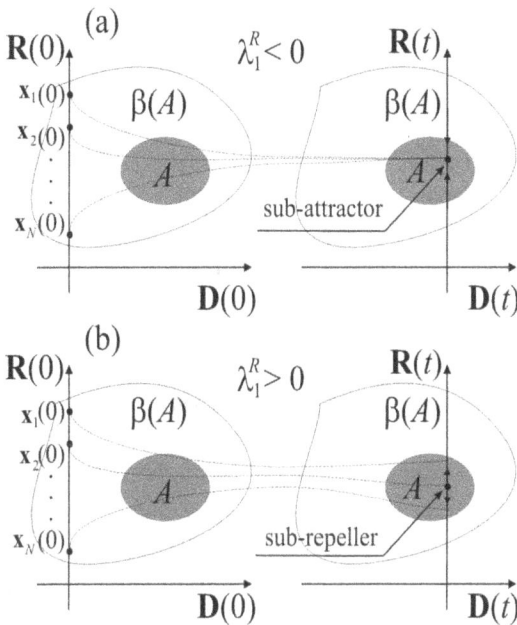

Fig. 3.7. Two-dimensional visualization of the mechanism of synchronization (a) and desynchronization (b); A – system attractor; $\beta(A)$ – basin of attraction of the attractor A; **D** – driving subspace; **R** – response subspace.

If at least one RLE is positive, then the synchronization between the oscillators under consideration is impossible because the instability associated with the positive RLE causes a divergence of nearby trajectories (Fig. 3.7b) in the *response subspace* and the *sub-attractor* becomes *sub-repeller*.

Chapter 4

Synchronizability of Coupled Oscillators

One of the most crucial features of any sets, arrays or networks of coupled dynamical systems is their so-called synchronizability. In general, this property can be quantified by a magnitude of synchronization thresholds determining these ranges and a total size of the synchronous range(s) in the overall coupling parameter space. In this Chapter the synchronization tendency in networks and arrays of oscillators with various types of the coupling is studied.

4.1 Single Synchronous Ranges of the Coupling Parameter

In most cases of coupled oscillators, a single synchronization range of the coupling parameter dominates. Such a range can be bottom-limited (see Figs. 4.8 and 4.9a), i.e., the synchronous state is stable after passing the critical limit of the increasing coupling parameter. On the other hand, it can be double-limited (Fig. 4.9b), i.e., there exists one window of synchronization (an interval of the coupling parameter) in the overall desynchronous regime (Fujisaka & Yamada (1983a), (1983b), Dmitriev *et al.* (1997), Pecora (1998), Fink *et al.* (2003), Barahona & Pecora (2002), Nishikawa *et al.* (2003), Stefanski *et al.* (2004)).

4.1.1 *Diagonally coupled systems*

Here, we present some results of analytical and numerical estimation of the synchronization threshold for chosen models of chaotic networks with the CD coupling between the oscillators. The analysis of synchronization stability on the basis of the DSSM has been compared to the results of the numerical experiment for a number of arrays with a regular structure of the coupling and for the networks with a random coupling configuration. In the numerical simulations, examples of classical dynamical systems (flows and maps) have been applied as the network nodes. Table 4.1 presents a form of the detailed equations which describe these examples with their corresponding LLEs.

Table 4.1. Dynamical systems used in the numerical simulations.

Dynamical system	Equations of motion	LLE — λ_1
Lorenz system	$dx/dt = 10(y - x)$ $dy/dt = - xz + 197x - y$ $dz/dt = xy - 8/3z$	1.849
Rössler system	$dx/dt = -y - z$ $dy/dt = x + 0.15y$ $dz/dt = 0.2 + z(x - 10)$	0.085
Duffing oscillator	$dx/dt = y$ $dy/dt = -x^3 - 0.1y + 10\sin(t)$	0.098
logistic map	$x_{n+1} = 3.9x_n(1 - x_n)$	0.485
Henon map	$x_{n+1} = 1 - 1.4x_n^2 + y_n$ $y_{n+1} = 0.3x_n$	0.419

4.1.1.1 *Regular coupling configuration*

In our analysis three cases of the arrays of chaotic systems with a regular structure of the coupling have been considered:

1. a *symmetrical global coupling* (each to each),
2. a star-type configuration of the coupling in three versions,
3. a chain-type (nearest neighbor) configuration.

In the case of a symmetrical global interaction, the connectivity matrix **G** and both DSSMs for flows (Eq. (3.27)) and for maps (Eq. (3.34)), respectively, assume the forms:

$$\mathbf{G} = d \begin{bmatrix} -N+1 & 1 & \cdots & 1 \\ 1 & \ddots & \ddots & \vdots \\ \vdots & \ddots & \ddots & 1 \\ 1 & \cdots & 1 & -N+1 \end{bmatrix}, \tag{4.1}$$

$$\mathbf{S} = \begin{bmatrix} \lambda_1 - Nd & 0 & \cdots & 0 \\ 0 & \lambda_1 - Nd & \cdots & 0 \\ \vdots & \vdots & \ddots & \vdots \\ 0 & 0 & \cdots & \lambda_1 - Nd \end{bmatrix} \tag{4.2}$$

and

$$\mathbf{M} = \exp(\lambda_1) \begin{bmatrix} 1 - Nd & 0 & \cdots & 0 \\ 0 & 1 - Nd & \cdots & 0 \\ \vdots & \vdots & \ddots & \vdots \\ 0 & 0 & \cdots & 1 - Nd \end{bmatrix}. \tag{4.3}$$

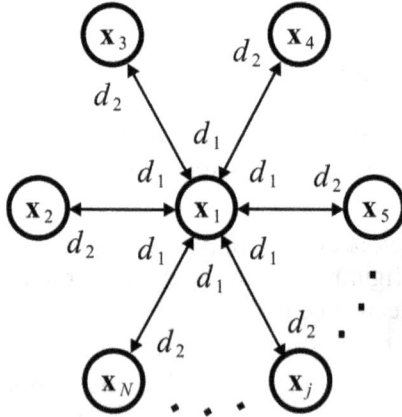

Fig. 4.1. Bidirectional star-type configuration of the coupling.

Inequalities (3.28) and (3.35) imply that the CS of all network nodes occurs in the following ranges of the coupling parameter:

$$d > \frac{\lambda_1}{N} \tag{4.4}$$

for flows, and

$$\frac{1 - \exp(-\lambda_1)}{N} < d < \frac{1 + \exp(-\lambda_1)}{N} \tag{4.5}$$

for maps. The above conditions of synchronization remain in agreement with the results obtained earlier by means of other approaches (Fujisaka & Yamada (1983a) and (1983b), Pikovsky (1984), Dmitriev *et al.* (1997), Stefanski (2000), Stefanski & Kapitaniak (2000), (2003b)).

A star-type configuration of the coupling (Fig. 4.1) can be realized in three versions, i.e., a mutual interaction (version I) and a *unidirectional coupling* to the central node (version II) or from it (version III). If we assume the first oscillator to be the central one in a star-type configuration of the coupling, then it is described by the equations:

$$\dot{\mathbf{x}}_1 = \mathbf{f}(\mathbf{x}_1) + \sum_{j=2}^{N} d_1 \mathbf{I}_k (\mathbf{x}_j - \mathbf{x}_1),$$

$$\dot{\mathbf{x}}_j = \mathbf{f}(\mathbf{x}_j) + d_2 \mathbf{I}_k (\mathbf{x}_1 - \mathbf{x}_j), \tag{4.6}$$

or

$$\mathbf{x}_{n+1}^1 = \mathbf{f}(\mathbf{x}_n^1) + \sum_{j=2}^{N} d_1 \mathbf{I}_k \left[\mathbf{f}(\mathbf{x}_n^j) - \mathbf{f}(\mathbf{x}_n^1) \right],$$

$$\mathbf{x}_{n+1}^j = \mathbf{f}(\mathbf{x}_n^j) + d_2 \mathbf{I}_k \left[\mathbf{f}(\mathbf{x}_n^1) - \mathbf{f}(\mathbf{x}_n^j) \right]. \tag{4.7}$$

The connectivity matrix and both DSSM corresponding to the systems given by Eqs. (4.6) and (4.7) are as follows:

$$\mathbf{G} = \begin{bmatrix} (1-N)d_1 & d_1 & \cdots & d_1 \\ d_2 & -d_2 & \ddots & 0 \\ \vdots & & \ddots & \ddots & \vdots \\ d_2 & 0 & \cdots & -d_2 \end{bmatrix}, \tag{4.8}$$

$$\mathbf{S} = \begin{bmatrix} \lambda_1 - (d_1 + d_2) & -d_1 & \cdots & -d_1 \\ -d_1 & \lambda_1 - (d_1 + d_2) & \cdots & -d_1 \\ \vdots & \vdots & \ddots & \vdots \\ -d_1 & -d_1 & \cdots & \lambda_1 - (d_1 + d_2) \end{bmatrix}, \quad (4.9)$$

$$\mathbf{M} = \exp(\lambda_1) \begin{bmatrix} 1 - (d_1 + d_2) & -d_1 & \cdots & -d_1 \\ -d_1 & 1 - (d_1 + d_2) & \cdots & -d_1 \\ \vdots & \vdots & \ddots & \vdots \\ -d_1 & -d_1 & \cdots & 1 - (d_1 + d_2) \end{bmatrix}. \quad (4.10)$$

The eigenvalues of both DSSM (Eqs. (4.9) and (4.10)) can be calculated analytically from the equations:

$$\left| \mathbf{S} - s\mathbf{I}_{N-1} \right| = \left[\lambda_1 - (N-1)d_1 - d_2 - s \right] (\lambda_1 - d_2 - s)^{N-2} = 0 \quad (4.11)$$

or

$$\left| \mathbf{M} - \mu\mathbf{I}_{N-1} \right| =$$
$$\left[\exp(\lambda_1)(1 - (N-1)d_1 - d_2) - \mu \right] \left[\exp(\lambda_1)(1 - d_2) - \mu \right]^{N-2} = 0 \quad (4.12)$$

The synchronization threshold of time-continuous systems for the first ($d_1 = d_2 = d$) and the third ($d_1 = 0$, $d_2 = d$) version of the star-type coupling configurations is given by the inequality:

$$d > \lambda_1. \quad (4.13)$$

The second version of the coupling structure ($d_1 = d$, $d_2 = 0$) allows the CS of periodic oscillators only, because the condition of synchronization (for flows and maps) resulting from Eqs. (4.11) and (4.12) assumes the form:

$$\lambda_1 < 0. \quad (4.14)$$

The next condition of synchronization, for the third version of maps coupled as a star ($d_1 = 0$, $d_2 = d$), is given by:

$$1 - \exp(-\lambda_1) < d < 1 + \exp(-\lambda_1). \quad (4.15)$$

The most interesting situation takes place when the *mutual coupling* in a star-type configuration of discrete-time systems is realized

($d_1=d_2=d$). Namely, the CS is guaranteed if the coupling coefficient d fulfills inequalities (4.5) and (4.15) simultaneously, i.e.,

$$1-\exp(-\lambda_1) < d < \frac{1+\exp(-\lambda_1)}{n}. \tag{4.16}$$

Thus, in such a case the maximum number of chaotic maps, which are able to synchronize, is limited by the inequality:

$$n < \frac{1+\exp(-\lambda_1)}{1-\exp(-\lambda_1)}. \tag{4.17}$$

The above presented synchronization ranges of the coupling parameter (inequalities (4.4), (4.5), (4.13) – (4.16)), which have been determined analytically, can be easily confirmed in numerical simulations with an arbitrary chaotic system assumed as the network node.

The last of the above-considered cases of the regular connection of nodes is a chain-type coupling configuration (*diffusive coupling*), where every oscillator interacts with two nearest neighbors (Fig. 4.2). The equations of motion for such a case are:

$$\dot{\mathbf{x}}_i = \mathbf{f}(\mathbf{x}_i) + d\mathbf{I}_k (\mathbf{x}_{i-1} - \mathbf{x}_i) + d\mathbf{I}_k (\mathbf{x}_{i+1} - \mathbf{x}_i), \tag{4.18}$$

and

$$\mathbf{x}_{n+1}^i = \mathbf{f}(\mathbf{x}_n^i) + d\mathbf{I}_k \left[\mathbf{f}(\mathbf{x}_n^{i-1}) - \mathbf{f}(\mathbf{x}_n^i) \right] + d\mathbf{I}_k \left[\mathbf{f}(\mathbf{x}_n^{i+1}) - \mathbf{f}(\mathbf{x}_n^i) \right]. \tag{4.19}$$

Hence, the connectivity configuration is defined by:

$$\mathbf{G} = d \begin{bmatrix} -2 & 1 & 0 & \cdots & 0 & 1 \\ 1 & -2 & 1 & \ddots & \cdots & 0 \\ 0 & 1 & \ddots & \ddots & \ddots & \vdots \\ \vdots & \ddots & \ddots & \ddots & 1 & 0 \\ 0 & \cdots & \ddots & 1 & -2 & 1 \\ 1 & 0 & \cdots & 0 & 1 & -2 \end{bmatrix}. \tag{4.20}$$

From the connectivity matrix (Eq. (4.20)), we obtain the following DSSMs:

$$S = \begin{bmatrix} \lambda_1 - 3d & d & 0 & \cdots & \cdots & 0 & -d \\ 0 & \lambda_1 - 2d & d & \ddots & \vdots & \vdots & \vdots \\ -d & d & \lambda_1 - 2d & \ddots & 0 & \vdots & \vdots \\ \vdots & 0 & d & \ddots & d & 0 & \vdots \\ \vdots & \vdots & 0 & \ddots & \lambda_1 - 2d & d & -d \\ \vdots & \vdots & \vdots & \ddots & d & \lambda_1 - 2d & 0 \\ -d & 0 & \cdots & \cdots & 0 & d & \lambda_1 - 3d \end{bmatrix}, \quad (4.21)$$

$$M = \exp(\lambda_1) \begin{bmatrix} 1 - 3d & d & 0 & \cdots & \cdots & 0 & -d \\ 0 & 1 - 2d & d & \ddots & \vdots & \vdots & \vdots \\ -d & d & 1 - 2d & \ddots & 0 & \vdots & \vdots \\ \vdots & 0 & d & \ddots & d & 0 & \vdots \\ \vdots & \vdots & 0 & \ddots & 1 - 2d & d & -d \\ \vdots & \vdots & \vdots & \ddots & d & 1 - 2d & 0 \\ -d & 0 & \cdots & \cdots & 0 & d & 1 - 3d \end{bmatrix}. \quad (4.22)$$

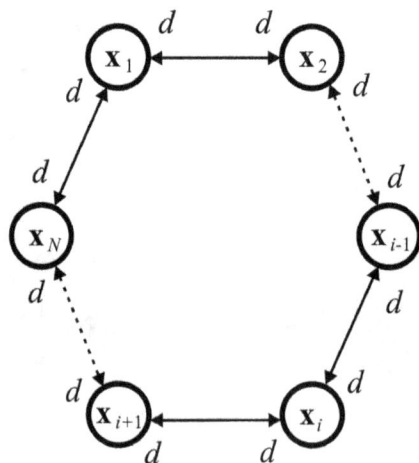

Fig. 4.2. Chain-type configuration of the coupling.

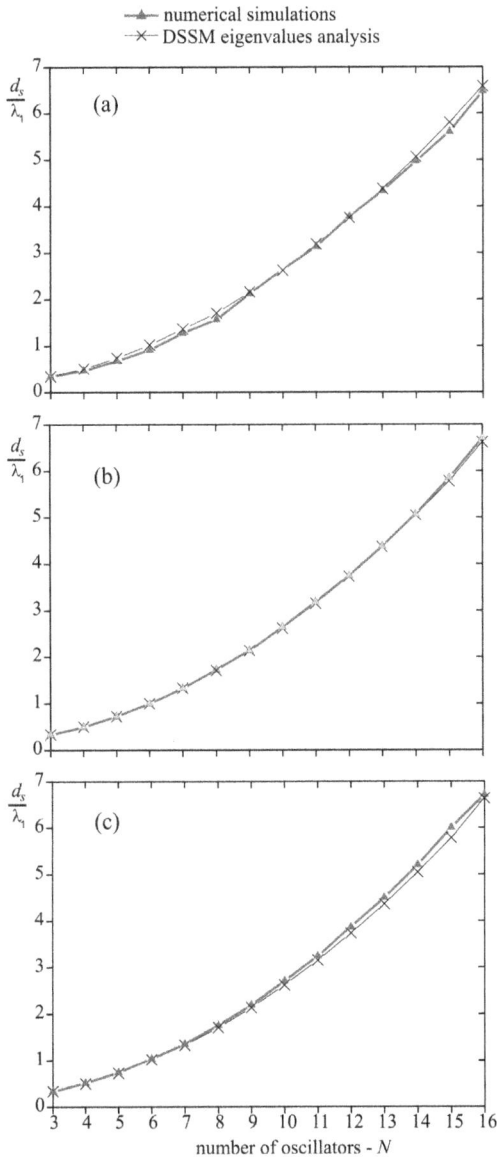

Fig. 4.3. Comparison of the synchronization threshold (the ratio d_s/λ_1 versus the number of oscillators in chain) calculated from the DSSM eigenvalues analysis and obtained from the numerical investigations of the chain synchronization; d_s — synchronization value of the coupling coefficient; (a) — Duffing oscillators, (b) — Lorenz systems, (c) — Rössler systems.

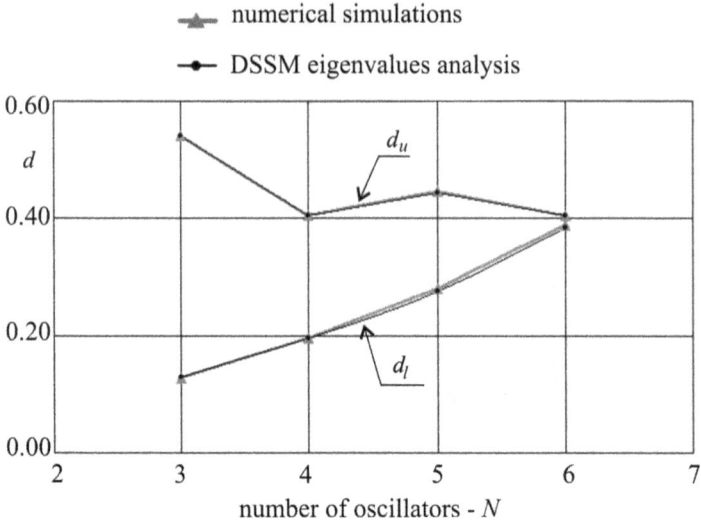

Fig. 4.4. Comparison of the synchronization ranges of the coupling coefficient d in the chain of diffusively coupled logistic maps calculated from the DSSM eigenvalues analysis and obtained from the numerical simulations; d_u and d_l — upper and lower ends of the synchronization range.

The synchronization threshold given by inequalities (3.28) and (3.35) has been evaluated numerically on the basis of the QR algorithm of eigenvalues calculation (Vetterling *et al.* (1992)). In Figs. 4.3a – 4.3c, a comparison of the results obtained from the DSSM eigenvalue analysis (Eq. (4.21)) and the from direct investigation of the synchronization process for time-continuous chaotic systems presented in Table 4.1 is illustrated. We can observe a high conformity level of the results in all three examples. A similar situation is shown in Fig. 4.4, where the synchronization analysis of the logistic map (from Table 4.1) chain is presented. Also, in this case the synchronization ranges of the coupling coefficient determined from the eigenvalues of DSSM (Eq. (4.22)) agree with the appropriate regions obtained from the numerical simulations of chain dynamics. Our analysis additionally demonstrates that for $N \geq 7$ the CS in the chain under consideration is impossible because for an arbitrary d, the condition of synchronization (inequality (3.35)) is not fulfilled.

4.1.1.2 *Random coupling configuration*

The first example consists of four randomly coupled chaotic oscillators according to the scheme shown in Fig. 4.5. The corresponding connectivity matrix has the following form:

$$\mathbf{G} = d \begin{bmatrix} -3 & 1 & 2 & 0 \\ 2 & -2 & 0 & 0 \\ 1 & 0 & -1 & 0 \\ 3 & 2 & 0 & -5 \end{bmatrix}. \tag{4.23}$$

It is obvious that an irregular coupling configuration causes a nonsymmetrical, random structure of both DSSM:

$$\mathbf{S} = \begin{bmatrix} \lambda_1 - 3d & -2d & 0 \\ -d & \lambda_1 - 3d & 0 \\ 0 & -2d & \lambda_1 - 4d \end{bmatrix}, \tag{4.24}$$

and

$$\mathbf{M} = \exp(\lambda_1) \begin{bmatrix} 1 - 3d & -2d & 0 \\ -d & 1 - 3d & 0 \\ 0 & -2d & 1 - 4d \end{bmatrix}. \tag{4.25}$$

The eigenvalues of the above matrices (Eqs. (4.24) and (4.25)) can be calculated analytically as:

$$s_1 = \lambda_1 - 4d, \; s_{2,3} = \lambda_1 - (3 \pm \sqrt{2}) d \quad \text{and}$$

$$\mu_1 = \exp(\lambda_1)(1 - 4d), \quad \mu_{2,3} = \exp(\lambda_1)[1 - (3 \pm \sqrt{2})d].$$

Substituting these eigenvalues into inequalities (3.28) and (3.35), we obtain the synchronization ranges of the parameter d for the network shown in Fig. 4.5:

$$d > \frac{\lambda_1}{3 - \sqrt{2}} \tag{4.26}$$

for flows, and

$$\frac{1 - \exp(-\lambda_1)}{3 - \sqrt{2}} < d < \frac{1 + \exp(-\lambda_1)}{3 + \sqrt{2}} \tag{4.27}$$

for maps. The confirmation of the CS stability regions given by inequalities (4.26) and (4.27) is shown in Figs. 4.6a and 4.6b, where a comparison of the analytical results with the numerical simulations is presented. As the examples of nodes in the considered network (Fig. 4.5), a Rössler oscillator and a Henon map (Henon (1976)) have been used.

The last example is a set of ten identical time-continuous systems (Duffing oscillators) with a randomly assumed connectivity structure represented by the matrix:

$$\mathbf{G} = d\begin{bmatrix} -17 & 0 & 2 & 4 & 0 & 0 & 7 & 0 & 4 & 0 \\ 0 & -17 & 2 & 8 & 0 & 1 & 0 & 0 & 0 & 6 \\ 0 & 0 & -17 & 1 & 5 & 0 & 7 & 4 & 0 & 0 \\ 9 & 2 & 0 & -21 & 0 & 3 & 5 & 1 & 0 & 1 \\ 4 & 6 & 7 & 1 & -23 & 0 & 0 & 3 & 2 & 0 \\ 5 & 0 & 0 & 0 & 10 & -19 & 3 & 0 & 0 & 1 \\ 0 & 1 & 3 & 10 & 2 & 0 & -18 & 0 & 2 & 0 \\ 0 & 4 & 3 & 0 & 0 & 5 & 7 & -19 & 0 & 0 \\ 1 & 0 & 0 & 2 & 0 & 0 & 0 & 10 & -13 & 0 \\ 7 & 4 & 0 & 0 & 10 & 2 & 0 & 1 & 2 & -26 \end{bmatrix} . (4.28)$$

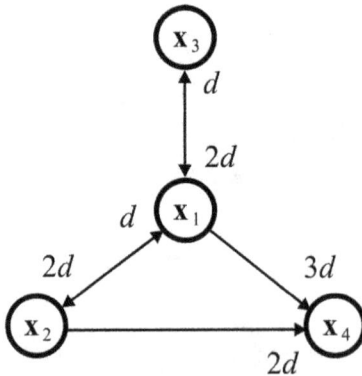

Fig. 4.5. Four oscillators with a irregular configuration of the coupling.

The DSSM resulting from matrix (4.28) is as follows:

$$\mathbf{S} = \lambda_1 \mathbf{I}_9 + d \begin{bmatrix} -17 & 0 & 4 & 0 & 1 & -7 & 0 & -4 & 6 \\ 0 & -19 & -2 & 5 & 0 & 0 & 4 & -4 & 0 \\ 2 & -2 & -25 & 0 & 3 & -2 & 1 & -4 & 1 \\ 6 & 5 & -3 & -23 & 0 & -7 & 3 & -2 & 0 \\ 0 & -2 & -4 & 10 & -19 & -4 & 0 & -4 & 1 \\ 1 & 1 & 6 & 2 & 0 & -25 & 0 & -2 & 0 \\ 4 & 1 & -4 & 0 & 5 & 0 & -19 & -4 & 0 \\ 0 & -2 & -2 & 0 & 0 & -7 & 10 & -17 & 0 \\ 4 & -2 & -4 & 10 & 2 & -7 & 1 & -2 & -26 \end{bmatrix} . \quad (4.29)$$

In Figs. 4.7a and 4.7b, numerically calculated real parts of the eigenvalues of the DSSM (Eq. (4.28), Fig. 4.7b) and the corresponding *trajectory separation (synchronization error)* bifurcation diagram (Fig. 4.7a) drawn as a function of the coupling parameter d are shown. Comparing both diagrams, we can see that according to inequality (3.28), the synchronization appears (*synchronization error* tends to zero in Fig. 4.7a) if all real parts of DSSM eigenvalues become negative. Thus, even in a case of a larger number of oscillators with a completely irregular coupling structure, the calculation of the synchronization threshold by means of the DSSM method is a simple task.

Summing up, the above theoretical analysis supported by the numerical simulations leads to the main conclusion that chaotic synchronization in the networks composed of identical oscillators with a diagonal, diffusive-type interaction between them can be considered as a simple, linear dynamical process. Two dominant factors, i.e., the largest Lyapunov exponent of the network node system and the effective coupling rate between the nodes, play the dominant role in this process. This property of the *diagonal coupling* allows us to estimate the synchronization threshold for an arbitrary configuration of the coupling. In order to examine the stability of the synchronization state, a concept of the DSSM has been applied. The DSSM is constructed directly from the coupling configuration matrix and allows for a linear stability analysis. Such a method is based on the simplified, linear model of the network. An advantage of this approach is simplicity of its application for both continuous-time and discrete-time systems.

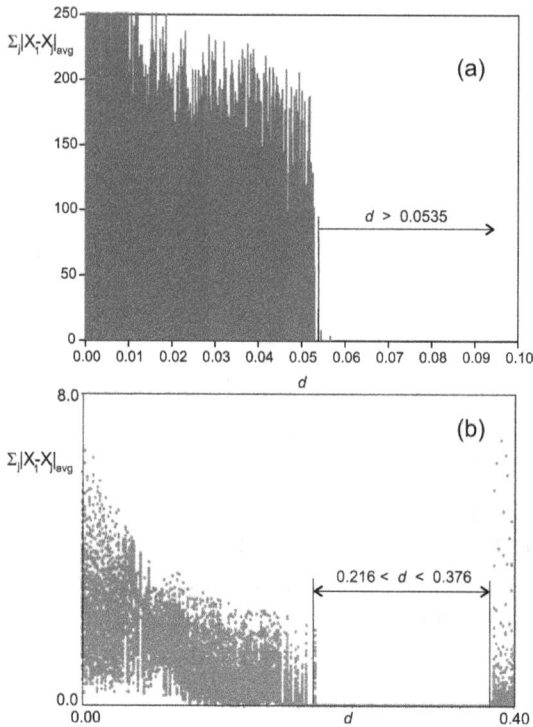

Fig. 4.6. Bifurcation diagrams (a sum of average *synchronization error* versus the coupling coefficient) representing a comparison of the synchronization ranges in the ensembles of dynamical systems with the scheme of connections shown in Fig. 4.5. The ranges obtained analytically according to Eqs. (4.26) and (4.27) are marked and described in black; (a) set of Rössler systems, (b) set of logistic maps.

However, one should note that our approach can be realized only in the case of the uniform (with identical diagonals) CD coupling because only then we can substitute the coupling matrices by the coupling coefficients according to Eq. (3.21b). The ND coupling forces us to take a full mathematical form of the node system in considerations of the

network synchronization process, which makes simplification of the
network given by Eqs. (3.21a) and (3.21b) impossible. In such cases
other techniques for determination of the synchronization condition have
to be used, for instance the MSF. Obviously, the MSF is a more general
approach than the DSSM. Hence, it also allows one to solve analytically
the synchronization threshold problem in CD-coupled systems. The MSF
of such a case is depicted in Fig. 4.8, where we can see a simple linear
relation between the LLE λ_1 and real eigenvalues of the connectivity
matrix α.

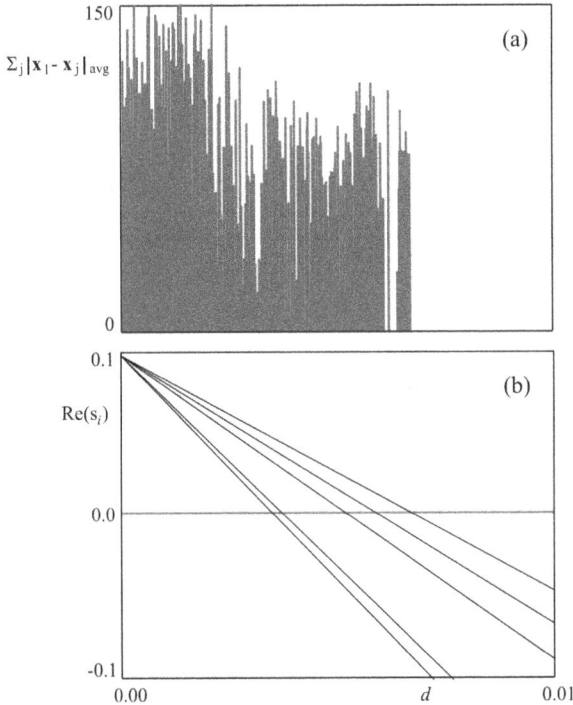

Fig. 4.7. Bifurcation diagram of average *synchronization error* versus the coupling
coefficient (a) and the corresponding eigenvalues of DSSM (Eq. (4.29)) (b) for ten
Duffing oscillators with a random structure of connections (Eq. (4.28)).

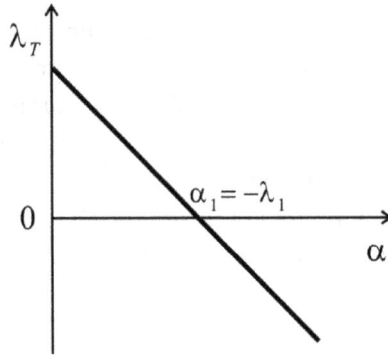

Fig. 4.8. MSF of diagonally coupled oscillators.

4.1.2 *Oscillators with the PD and ND coupling*

A type of the single synchronous range appearing in the systems with a PD coupling is depended on the CLEs of the remaining, uncoupled sub-block of the node system. This property results from the asymptotic effect of the PD coupling (Pecora *et al.* (2000), Fink *et al.* (2000)). An essence of this effect, depicted in Figs. 4.9a and 4.9b, is that the largest TLE (λ_T) tends asymptotically to the value of the largest CLE (λ_C) for a strong coupling. In order to explain it in more detail, let us consider again the classical Rössler system analyzed in Secs. 2.3 and 3.3.3:

$$\dot{x} = -y - z,$$
$$\dot{y} = x + ay, \qquad\qquad (4.30)$$
$$\dot{z} = b + z(x - c).$$

For instance, the PD coupling between the oscillators (4.30) can be applied via an x-component, i.e., with the output function as follows:

$$\mathbf{H} = \begin{pmatrix} 1 & 0 & 0 \\ 0 & 0 & 0 \\ 0 & 0 & 0 \end{pmatrix}. \qquad\qquad (4.31)$$

In such a situation, the coupling damps the x-perturbation to zero for the large negative α. Then, in the generic variational equation for calculating the MSF (Eq. (3.40)), the uncoupled sub-block yz of the Jacobian (see Eq. (3.55)) becomes to be dominated. Thus, the TLE must asymptote to an unstable level of a with the increasing coupling, i.e., it approaches the corresponding positive CLE (see Fig. 4.9b).

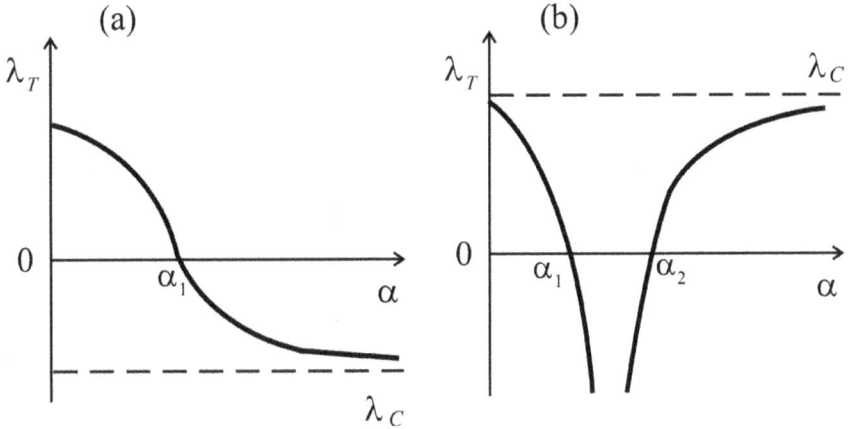

Fig. 4.9. Typical examples of the MSF — $\lambda_T(\alpha)$ by the *real coupling*: (a) bottom-limited synchronous range (α_1, ∞), (b) double-limited synchronous interval (α_1, α_2). Additionally, the asymptotic coupling effect is illustrated — λ_T converges to λ_C with the increasing coupling represented by the α

On the other hand, if a large PD y-coupling is introduced to connect Rössler systems (4.30), i.e.,

$$\mathbf{H} = \begin{pmatrix} 0 & 0 & 0 \\ 0 & 1 & 0 \\ 0 & 0 & 0 \end{pmatrix}, \tag{4.32}$$

then the Jacobian of the undamped xz sub-block (see Eq. (3.57)) is characterized by a pair of negative eigenvalues (Eq. (3.58)) approximating the CLEs. In that case the MSF tends to a larger negative eigenvalue when the coupling strength increases, similarly as shown in Fig. 4.9a. Therefore, for the negative λ_C of the uncoupled sub-block, the synchronous range is only bottom-limited (Fig. 4.9a) and for the positive λ_C such a range is double-limited (Fig. 4.9b).

In ND coupled systems or in networks with any other combined PD-ND linking functions, an existence of double and bottom-limited regions of synchronization is also possible. However, in such cases the asymptotic coupling effect illustrated in Figs. 4.9a and 4.9b usually does not occur. On the other hand, it can happen, especially for ND coupling schemes, that the synchronization will not appear for any coupling strength, i.e., the TLE is positive in the entire range of α.

4.1.2.1 *Bottom-limited synchronizability*

If the MSF providing a single range of the negative TLE is only bottom-limited, then the synchronous range of the coupling parameter σ is simply reflected from the synchronous α-range of the MSF only via the eigenvalue γ_1 determining the boundary (the smallest) value of σ, required for an appearance of synchronization. The mechanism of such a reflection is explained in Fig. 4.12. Here it should be noted that the first approach suggesting employing this property for an estimation of the network synchronizability is the Wu–Chua conjecture (Wu & Chua (1996)) quoted below.

Theorem 4.1 (Wu – Chua conjecture) *If the synchronization threshold of the M number of coupled oscillators amounts σ_M, then another network of similarly coupled systems has an analogous threshold σ_N, such that the following relation between them is fulfilled*

$$\sigma_M \gamma_{1(M)} = \sigma_N \gamma_{1(N)}, \qquad (4.33)$$

where $\gamma_{1(M)}$ and $\gamma_{1(N)}$ are the largest non-zero eigenvalues of the connectivity matrix **G** *for the networks of M and N oscillators, respectively.*

Conjecture 4.1 allows us to estimate the synchronization threshold σ_N in any size of the network using the synchronization threshold σ_2 for two coupled oscillators according to the formula. Since $\gamma_{1(2)} = -2$, we have:

$$\sigma_N = \frac{2\sigma_2}{\left|\gamma_{1(N)}\right|}, \qquad (4.34)$$

i.e., an equivalence of Eq. (3.41). Consequently, this approach can be treated as an analogy to the *two oscillators probe* (Eq. (3.53)).

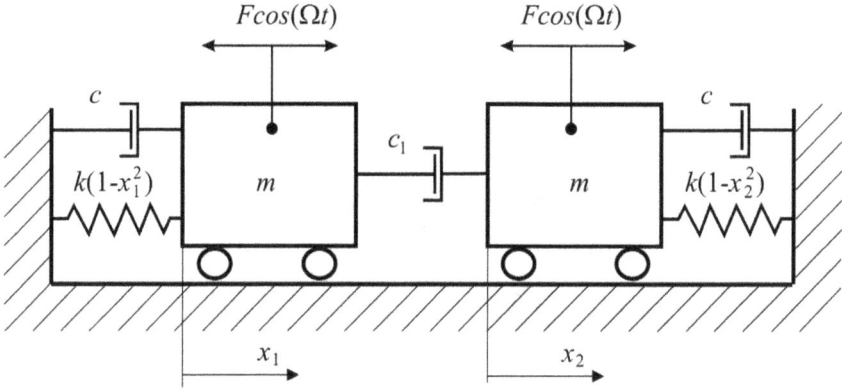

Fig. 4.10. Pair of non-linear mechanical oscillators of the Duffing type with a *dissipative coupling*.

In order to illustrate the considered strategy for predicting the bottom-limited synchronizability, let us focus on the real equivalent of the PD coupling, i.e., mechanical oscillators of the Duffing type with a *dissipative coupling* between them. The physical model of two coupled double-well Duffing systems, which can be used in the experimental *two oscillators probe* is shown in Fig. 4.10. This identical twin system is described with the following non-dimensional equations:

$$\begin{aligned}
\dot{x}_1 &= y_1, \\
\dot{y}_1 &= q\cos(\eta t) + ax_1(1 - x_1^2) - hy_1 + \sigma(y_2 - y_1), \\
\dot{x}_2 &= y_2, \\
\dot{y}_2 &= q\cos(\eta t) + ax_2(1 - x_2^2) - hy_2 + \sigma(y_1 - y_2),
\end{aligned} \tag{4.35}$$

where q, η, a, and h are dimensionless representations of the real parameters F, Ω, k and c, respectively. The coupling between vibrating masses is realized by a viscous damper with the damping coefficient c_1. which is represented by the dimensionless coupling parameter σ. Thus, the corresponding linking function is:

$$\mathbf{H} = \begin{pmatrix} 0 & 0 \\ 0 & 1 \end{pmatrix}.$$

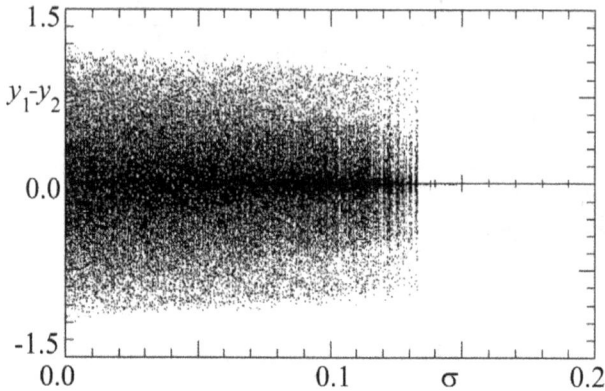

Fig. 4.11. Bifurcation diagrams of the *trajectory separation* versus the coupling coefficient σ for the PD coupling (Eq. (4.35)) of two chaotic Duffing oscillators; $q = 0.30$, $\eta = 1.00$ $a = 1.00$ and $h = 0.25$.

The bifurcation diagram (the *trajectory separation* versus the coupling coefficient) of system (4.35), where a value of the critical synchronous coupling strength $\sigma_2 = 0.135$ is clearly readable, is depicted in Fig. 4.11. The value of σ_2 can be applied as a result of *two oscillators probe*, e.g., in an application of the Wu – Chua conjecture to any network with the bottom-limited synchronous region.

Extend now twin system (4.35) into a chain of ten diffusively coupled oscillators,

$$\dot{x}_i = y_i,$$
$$\dot{y}_i = q\cos(\eta t) + ax_i(1 - x_i^2) - hy_i + \sigma(y_{i+1} + y_{i-1} - 2y_i), \qquad (4.36)$$

where $i = 1, 2, \ldots, 10$.

Figure 4.12 illustrates how the synchronous range of the MSF is reflected on the coupling parameter space according to Eq. (3.41) with an example of system (4.36). Here, we can see a projection of the negative TLE range via eigenvalues of the connectivity matrix to the bifurcation diagram of the *synchronization error* versus the coupling coefficient σ. The connection in the 1D array gives the eigenvalues of **G** defined by formula (2.12b), thus for $N = 10$ we have nine negative eigenvalues. Four of them are twice degenerated, so we can distinguish five different ones

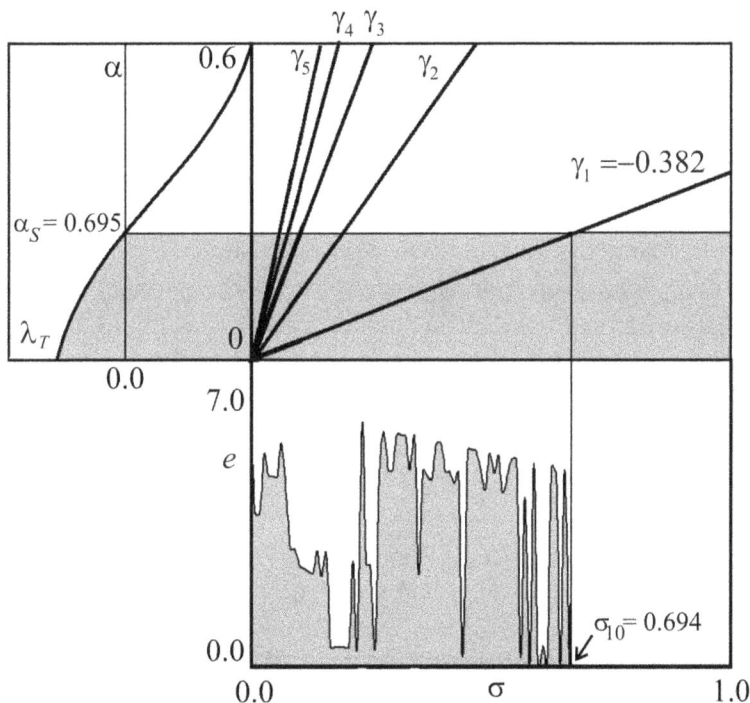

Fig. 4.12. Projection from the MSF $\lambda_T(\alpha)$ diagram, via eigenvalues γ_k of the connectivity matrix **G**, to the bifurcation diagram of the synchronization error e (computed according to Eq. (4.43) in the next Sec.) versus the coupling coefficient σ for an array of $N = 10$ Duffing oscillators (Eq. (4.36)). The desynchronous interval is depicted in gray. The CS takes place in σ-ranges where e approaches the zero value; $q = 0.30$, $\eta = 1.00$ $a = 1.00$ and $h = 0.25$.

between them, with the largest $\gamma_1 = -0.382$. The synchronization mechanism is shown in Fig. 4.12. A synchronous collective motion is possible if all the eigenvalues, representing transversal modes, can be found in the bottom-limited range of the negative TLE when the coupling strength σ increases. Thus, a dominant role of the largest eigenvalue of the connectivity matrix is clearly visible. From the MSF diagram in Fig. 4.12 it results that the smallest synchronous value of α is $\alpha_S = 0.265$. Hence, in accordance with Eq. (3.41), the synchronous value of the coupling coefficient amounts $\sigma_{10} = \alpha_S / |\gamma_1| = 0.694$. Almost the same synchronization threshold arises from the Wu–Chua conjecture

(Eq. (4.34)) for the value $2\sigma_2=0.270$ estimated from the *two oscillators probe*. This threshold is corroborated in the corresponding *synchronization error* diagram at the bottom of Fig. 4.12.

Moreover, from the above analysis, the next conjecture follows such that the synchronizability of not diagonally coupled systems can be also estimated using the DSSM technique. Namely, the *two oscillators probe* for the CD coupling gives a result $\sigma_2 = \lambda_1/2$. Thus, we can assume that substituting $2\sigma_2$, taken from a similar probe with any other kind of the coupling (different from the CD one), instead of λ_1 in Eq. (3.21a) allows a linear approach of the network according to the DSSM concept but with no diagonal connections. We have examined this conjecture for the considered example of ten Duffing oscillators (Eq. (4.36)) and we have obtained a satisfactorily convergence of the results with the others presented here.

4.1.2.2 *Double-limited synchronous interval*

For the case of the MSF with a double-limited α-interval of the negative TLE (Fig. 4.9b), two transverse eigenmodes have an influence on the σ-limits of the synchronous regime: the longest spatial-frequency mode, corresponding to the largest eigenvalue γ_1, and the shortest spatial-frequency mode, corresponding to the smallest eigenvalue γ_{N-1}. These both eigenvalues determine the width of the synchronous σ-range and two types of desynchronizing bifurcations can occur when the synchronous state looses its stability (see Figs. 4.13a and 4.13b). The decreasing σ leads to a *long wavelength bifurcation* (LWB), because the longest wavelength mode ξ_1 becomes unstable. On the other hand, the increasing coupling strength causes the shortest wavelength mode ξ_{N-1} to become unstable, thus a *short wavelength bifurcation* (SWB) takes place (Pecora (1998)). Another characteristic feature of coupled chaotic systems with a double-limited synchronous interval is the array size limit, i.e., the maximum number of oscillators in array which are able to synchronize. For the number of oscillators, which is larger than the size limit, the synchronous σ-interval does not exist. Such an interval exists if:

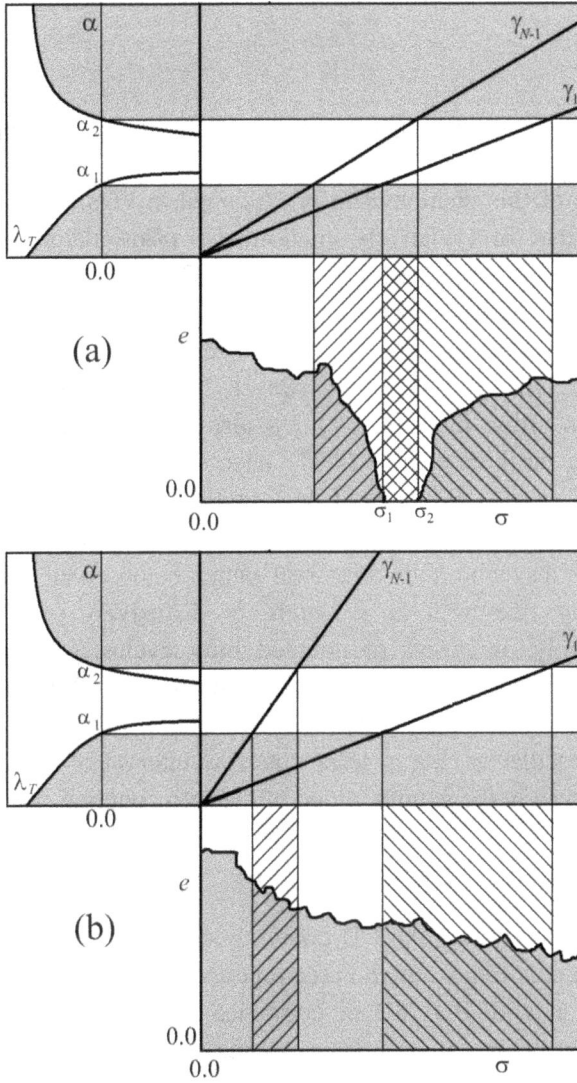

Fig. 4.13. Mechanism of the array size limit; (a) $\gamma_{N-1}/\gamma_1 < \alpha_2/\alpha_1$: synchronous intervals of the longest and shortest frequency modes (hatched) are partly overlapped and, as a result, a stable synchronous range of the coupling strength (σ_1, σ_2) exists; (b) $\gamma_{N-1}/\gamma_1 > \alpha_2/\alpha_1$: synchronous intervals of both boundary modes are disconnected, i.e., the CS of all oscillators in array is impossible. The desynchronous regions are depicted in gray.

$$\frac{\gamma_{N-1}}{\gamma_1} < \frac{\alpha_2}{\alpha_1}, \tag{4.37}$$

where α_1 and α_2 are the boundaries of the synchronous α-interval (see Fig. 4.9b).

A nature of this phenomenon is explained in Figs. 4.13a and 4.13b. We can see that for a relatively small number of oscillators in array, i.e., under the size limit defined by inequality (4.37) in each individual case, there exists a region of σ where both the maximum and minimum nonzero eigenvalues of the **G** matrix are located in the interval of the negative TLE (Fig. 4.13a). However, for an increasing number of oscillators in array, the ratio γ_{N-1}/γ_1 also grows (in accordance with Eq. (2.12b)) until inequality (4.37) becomes not fulfilled. Then, the eigenvalues γ_1 and γ_{N-1} cannot be found together in the synchronous interval of the MSF when the coupling strength increases (or decreases) and therefore a synchronous σ-interval cannot be observed (Fig. 4.13b).

The array size limit is a feature of diffusively coupled chaotic oscillators only. In arrays of coupled limit cycles, a double (upper) limited synchronous windows $(0, \sigma_1)$ can occur as well. Such a window will be strongly compressed down towards the origin for a growing number of oscillators, but at least a narrow interval of synchronization can be observed in the neighborhood of the zero coupling strength.

4.1.2.3 *Effectiveness of the PD coupling*

It seems intuitively that the CD coupling should be the most effective, i.e., the least coupling strength is required to achieve the synchronization in comparison with PD, ND or combined coupling functions. Such a conjecture results from a regular character of the convergence effect of the CD coupling (Eq. (3.20)) illustrated in Fig. 3.1b.

In order to verify a validity of this conjecture, consider two coupled identical Lorenz oscillators:

$$\dot{x}_1 = -\delta(x_1 - y_1) + d_x(x_2 - x_1),$$
$$\dot{y}_1 = -x_1 z_1 + r x_1 - y_1 + d_y(y_2 - y_1),$$
$$\dot{z}_1 = x_1 y_1 - b z_1 + d_z(z_2 - z_1),$$
$$\dot{x}_2 = -\delta(x_2 - y_2) + d_x(x_1 - x_2),$$
$$\dot{y}_2 = -x_2 z_2 + r x_2 - y_2 + d_y(y_1 - y_2),$$
$$\dot{z}_2 = x_2 y_2 - b z_2 + d_z(z_1 - z_2).$$

(4.38)

For $\delta = 10.0$, $b = 8/3$ and $r = 120.0$, the Lorenz system works in a chaotic regime characterized by the positive LLE ($\lambda_1 = 1.576$).

During the numerical simulations, three forms of the *symmetrical coupling* were considered:

1. a CD xyz-coupling ($d_x = d_y = d_z = d$),
2. a PD yz-coupling ($d_x = 0$, $d_y = d_z = d$),
3. a PD x-coupling ($d_x = d$, $d_y = d_z = 0$),
4. a PD y-coupling ($d_x = 0$, $d_y = d$, $d_z = 0$),
5. a PD z-coupling ($d_x = d_y = 0$, $d_z = d$).

The numerically generated synchronization values d_s are presented on the bifurcation diagrams shown in Figs. 4.14a, 4.14b and 4.14c. In accordance with the analytical approach given by inequality (3.17), the CS in system (4.38) appears when the coefficient of the symetrical CD coupling oversteps the critical value $d_s \approx \lambda_1/2$ (Fig. 4.14a). In the case of the PD z-coupling, the synchronous value is visibly larger than $\lambda_1/2$ (Fig. 4.14b). Thus, it seems that the above-mentioned conjecture on the best efectiveness of the CD coupling is confirmed. However, the bifurcational diagram in Fig. 4.14c shows that this conjecture is false. The PD yz-coupling leads to the synchronization of analyzed Lorenz systems (4.38) under the CD coupling critical level $\lambda_1/2$. Such an interesting effect can be graphically explained in a way shown in Figs. 4.15. Here, a comparison of x, y and z PD couplings in system (4.38) are presented in the form of bifurcational plots of the largest TLE versus the coupling coefficient in the same range as in Fig. 4.14.

$$\lambda_1 = 1.5762$$

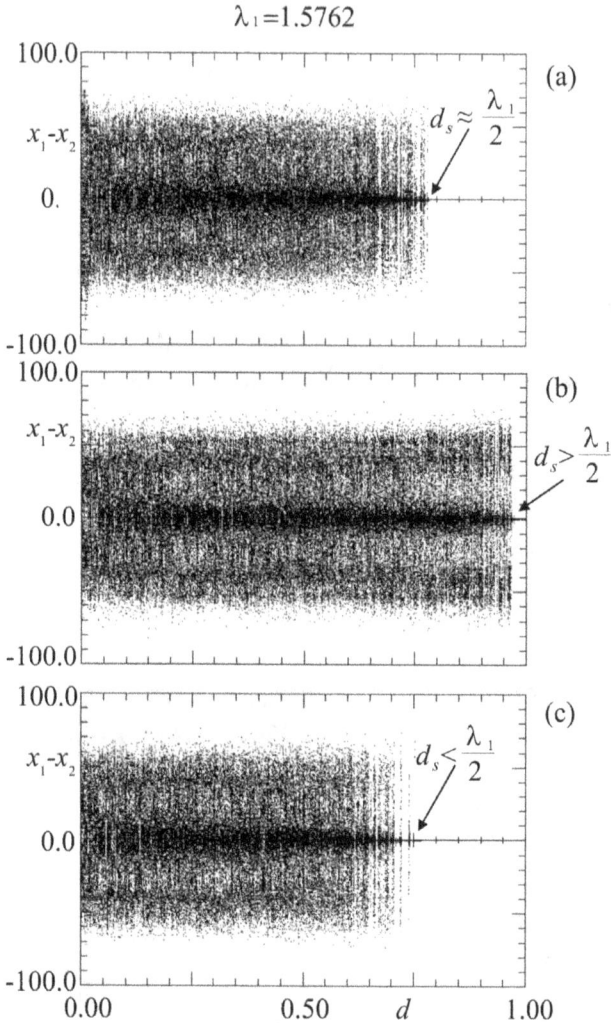

Fig. 4.14. Bifurcation diagrams of the *trajectory separation* versus the coupling coefficient d for coupled chaotic Lorenz systems (Eqs. (4.38)): the CD xyz-coupling, the PD yz-coupling, the PD z-coupling; $\sigma=10.0$, $b=8/3$, $r=120.0$. The corresponding value of the largest Lyapunov exponent — on the top.

In the case of the y ($d_x = 0$, $d_y = d$, $d_z = 0$) or z ($d_x = d_y = 0$, $d_z = d$) PD coupling (Fig. 4.15), we can observe a *convergence effect*, i.e., the largest TLE decreases with an increase in the coupling. On the other hand, the x-coupling causes a small increase of the largest TLE in comparison with the LLE of the separated Lorenz oscillator λ_1. Hence, the trajectories representing the subsystems of Eq. (4.38) rather diverge than converge in the phase space if only the x-coupling is introduced. To conclude, a lack of the x-component in the diagonal scheme of the non-intensive coupling leads to better synchronizability than in the case of the CD coupling because then only a superposition of the y and z coupling influences the synchroization process (Stefanski & Kapitaniak (2003c)).

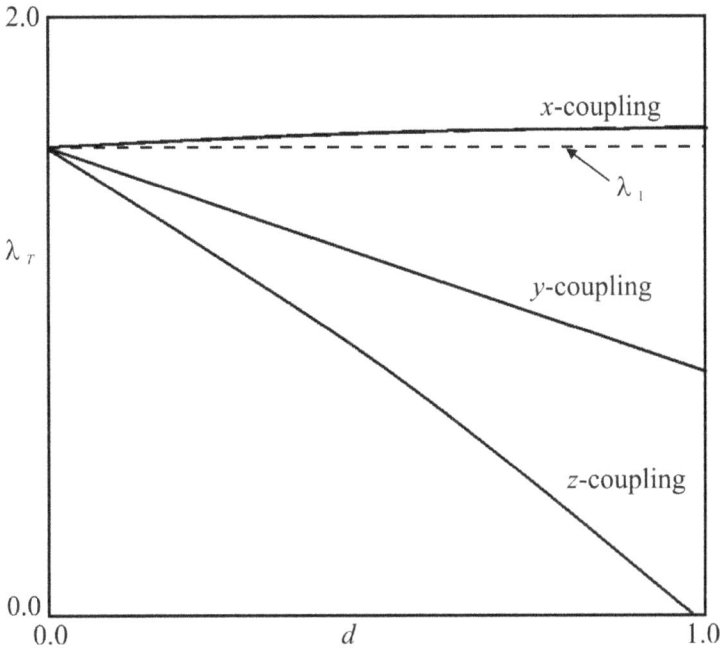

Fig. 4.15. Bifurcation diagrams of the TLE versus a small coupling coefficient d corresponding to Figs. 4.14a, 4.14b and 4.14c for the x ($d_x = d$, $d_y = 0$, $d_z = 0$), y ($d_x = 0$, $d_y = d$, $d_z = 0$) and z ($d_x = d_y = 0$, $d_z = d$) PD coupling in system (4.38).

4.2 Ragged Synchronizability

Apart from the cases analyzed in the previous Sec., there can exist more than one separated range of synchronization when the coupling strength increases. Then, an appearance or disappearance of desynchronous windows in the coupling parameter space can be observed, when a number of oscillators in the array or a topology of connections changes. This phenomenon has been called *ragged synchronizability* (Stefanski *et al.* (2004)). Similar effect has been also shown in Refs. (Yanchuk & Maistrenko (2003), Chen G. *et al.* (2008)). In this Sec., a mechanism governing this effect is explained and its influence on the global network dynamics is analyzed.

4.2.1 *Disconnected synchronous intervals*

In the numerical analysis, a classical single-well Duffing oscillator:

$$\ddot{x} + h\dot{x} + x^3 = q\sin(\eta t), \tag{4.39}$$

has been applied as an array node. The motion of each oscillator coupled in the array is governed by the following first-order differential equations:

$$\begin{aligned}
\dot{x}_i &= y_i, \\
\dot{y}_i &= -x_i^3 - hy_i + q\sin(\eta t) + \sigma(x_{i+1} + x_{i-1} - 2x_i),
\end{aligned} \tag{4.40}$$

where q, η h are the amplitude and the frequency of the harmonic forcing and the damping coefficient, respectively, $i = 1, 2, \ldots , N$. In the numerical analysis, we have assumed q as a control parameter and the following constant values: $\eta = 1.0$ and $h = 0.1$. Equation (4.40) models a chain of non-linear mechanical oscillators coupled with linear springs of the dimensionless stiffness σ (see Fig. 2.2b). Such a connection of oscillators can be classified as a case of the ND coupling due to the form of the output function:

$$\mathbf{H} = \begin{pmatrix} 0 & 0 \\ 1 & 0 \end{pmatrix}. \tag{4.41}$$

The structure of the nearest-neighbor connections of array nodes is described by the symmetrical connectivity matrix (Eq. (2.10)) with real eigenvalues (Eq. (2.12b)). Substituting the analyzed system (4.40) into Eq. (3.40), we obtain a generic variational equation for calculating the MSF, i.e., $\lambda_T(\alpha)$, in the form:

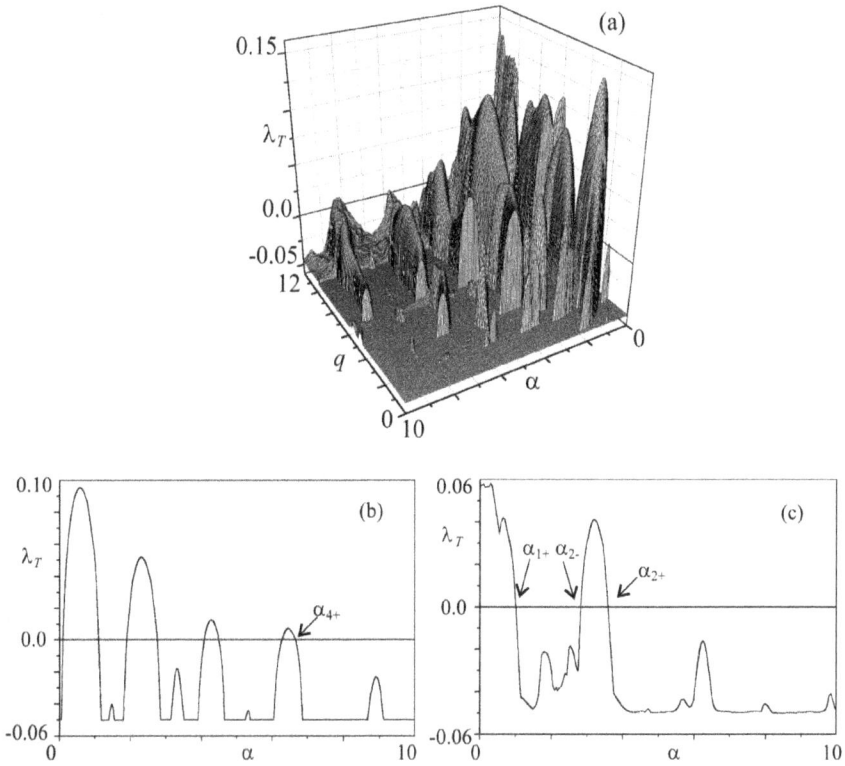

Fig. 4.16. Bifurcation diagram of the MSF $\lambda_T(\alpha)$ versus the amplitude of forcing q (a) and its cross-sections: q=7.0 (b), q=5.6 (c).

$$\dot{\xi} = \psi,$$
$$\dot{\psi} = -3ax^2\xi - h\psi + \alpha\xi. \tag{4.42}$$

The 3D diagram $\lambda_T(\alpha, q)$ shown in Fig. 4.16a can be treated as a bifurcation diagram of the MSF $\lambda_T(\alpha)$ versus the amplitude of forcing q, calculated for the system under consideration (Eqs. (4.40) and (4.42)). On the cross-sections of the MSF surface for isolated (i.e., when α=0) node systems working in the periodic (Fig. 4.16b) or chaotic (Fig. 4.16c) regime, characteristic folds or bubbles are visible. If such a bubble or fold emerges over the α axis, then a desynchronous interval of the coupling parameter σ appears and alternately appearing "windows" of synchronization and desynchronization can be observed, before the final synchronous state is achieved due to the increasing coupling strength. We introduce the term *ragged synchronizability* in order to describe this phenomenon.

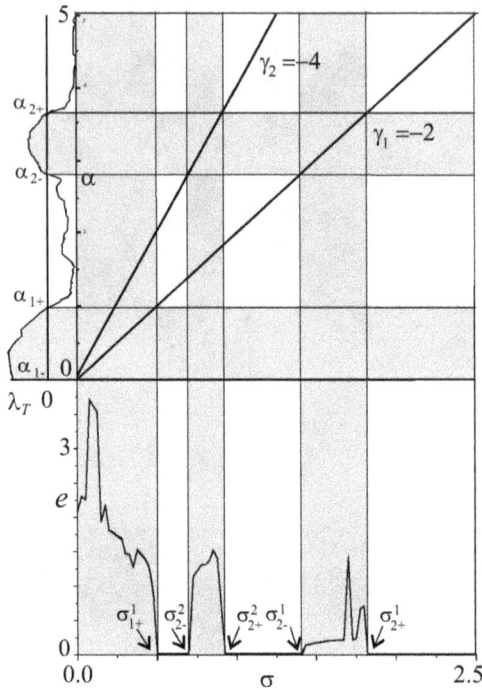

Fig. 4.17. Desynchronizing mechanism: a projection from the MSF $\lambda_T(\alpha)$ diagram, via eigenvalues γ_k of the connectivity matrix **G**, to the bifurcation diagram of the synchronization error e versus the coupling coefficient σ for an array of $N=4$ oscillators, $q=5.6$. The desynchronous intervals are in gray. The complete synchronization takes place in the σ-ranges where e approaches the zero value.

However, desynchronous intervals of the coupling parameter σ not always are a direct reflection of the α-intervals, where the largest TLE is positive. In networks of coupled chaotic oscillators exhibiting the folding MSF, the desynchronizing scenario leading to the *ragged synchronizability* can be more complicated. In order to explain this mechanism, consider a regular array (ring) of four oscillators (Eq. (4.40)) working in the chaotic regime, when uncoupled, with the MSF shown n Fig. 4.16c. Looking at the MSF diagram, we can distinguish two synchronous ranges: a double-limited window of synchronization $(\alpha_{1+}, \alpha_{2-})$ and a bottom-limited final synchronous range (α_{2+}, ∞). In accordance with Eq. (2.12b), for $N=4$ we have two non-zero eigenvalues of the connectivity matrix $\mathbf{G}$: $\gamma_1 = \gamma_3 = -2$ (twice degenerated) and $\gamma_2 = -4$. The desynchronizing mechanism is explained in Fig. 4.17, where a projection from the MSF diagram on the bifurcation diagram of the synchronization error:

$$e = \sum_{i=2}^{N} \sqrt{(x_1 - x_i)^2 + (y_1 - y_i)^2} \, , \qquad (4.43)$$

versus the coupling strength σ is shown. We can observe the third, intermediate desynchronous σ-interval $(\sigma^2_{2-}, \sigma^2_{2+})$ in comparison with only two desynchronous α-ranges (a meaning of the lower and upper indices is explained below). This interval appears as a result of the mode 2 desynchronizing bifurcation. Mode 2 crosses the second desynchronous α-interval $(\alpha_{2-}, \alpha_{2+})$, while mode 1 is still located in the first synchronous α-interval $(\alpha_{1+}, \alpha_{2-})$ and two synchronous windows $(\sigma^1_{1+}, \sigma^2_{2-})$ and $(\sigma^2_{2+}, \sigma^1_{2-})$ can be observed instead of the only one $(\sigma^1_{1+}, \sigma^1_{2-})$.

On the other hand, for a different number of oscillators in the array or for the case of the network of different topology of connections, the additional desynchronous interval may not appear. An existence of an intermediate window of desynchronization for a varying number of oscillators is depicted in Fig. 4.18, where all the σ-ranges correspond to the same α-range, in order to simplify the comparison. We can see that an increasing number of oscillators in the array produces consecutive transverse modes. However, only two of them (the longest ones) have a significant influence on the existence of desynchronous intervals (represented by bubbles in Fig. 4.18). The second mode desynchronous window approaches and then meets the first desynchronous window of mode 1, when the number of oscillators increases. For $N \geq 10$ it is completely consumed by the first mode window. It guarantees that only mode 1 associated with the largest eigenvalue γ_1 determines the ranges of synchronization, and thus the σ-intervals are a direct reflection of the

MSF according to Eq. (3.41). The same situation takes place for N=2, 3, because then only one nonzero eigenvalue of **G** exists.

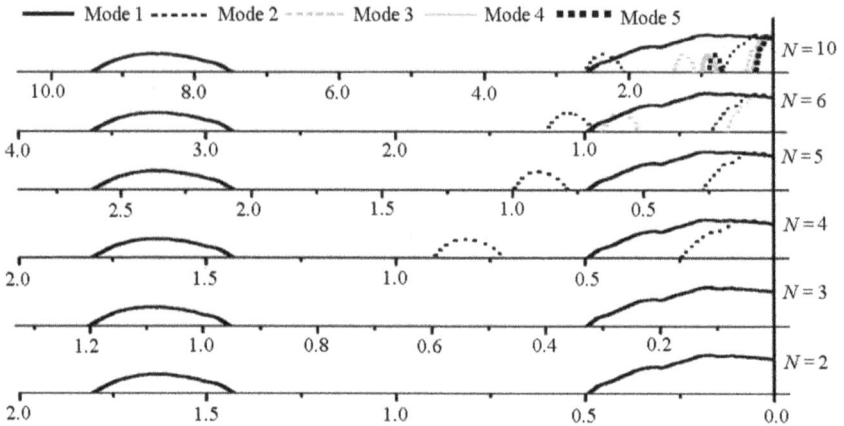

Fig. 4.18. Desynchronous σ-intervals (bubbles) associated with different transverse modes for a varying number N of oscillators in the array, q=5.6. All the σ-ranges correspond to the interval $0 < \alpha < 4.0$.

The example described above shows that for node systems having a MSF with multiple synchronous intervals, desynchronizing bifurcations of different modes are possible. This is in contrast to the systems with a single synchronous range where such bifurcations are determined at most by two eigenmodes, as we have mentioned in Sec. 2. Therefore, we have introduced some special notation for the boundaries between synchronous and desynchronous ranges of coupling strength. By the symbols α_{j-}, α_{j+} we have indicated the borders of consecutive desynchronous ranges of the MSF, $j = 1, 2, ..., s$, where s is the number of desynchronous α-ranges. The signs "+" or "–" in the subscript correspond to the lower (–) and higher (+) boundaries of these ranges. Consequently, the symbols of the borders of desynchronous ranges of coupling strength are σ^k_{j-}, σ^k_{j+}. The proposed notation is depicted in Figs. 4.16, 4.17 and 4.20. It is clearly visible that the characters k, $j-$ or $j+$ result from an intersection of the lines representing the boundary values α_{j-}, α_{j+} with a slope representing the eigenvalue γ_k. This notation shows which mode desynchronizing bifurcation (superscript) takes place during the transition from the synchronous to desynchronous regime and which desynchronous interval of the MSF (subscript) is associated with the given boundary value of the coupling coefficient.

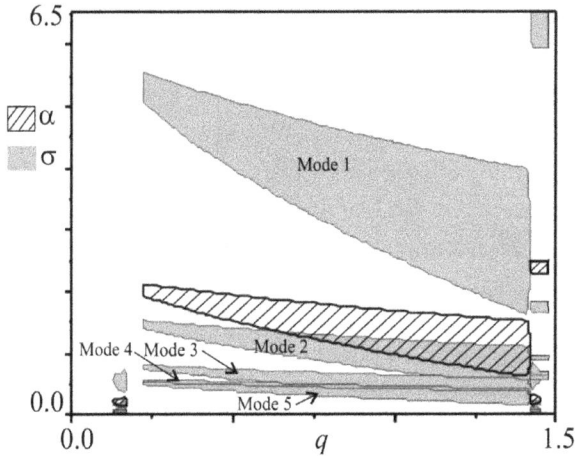

Fig. 4.19. Desynchronous α-region (hatched) and the corresponding desynchronous σ-regions (in gray) associated with five different transverse modes, for an array of $N=10$ oscillators working (when uncoupled) in the periodic range of the parameter q.

Another interesting *ragged synchronizability* effect can be observed for the array of periodic (when separated) oscillators under consideration (Eq. (4.40)). In such a case, the first α-interval $(0, \alpha_{1-})$ corresponds to the synchronized state due to the initially negative TLE. Increasing α, we can observe one or more desynchronous windows due to "bubbles" of the positive TLE (Fig. 4.16b). In Fig. 4.19 the corresponding α and σ desynchronous regions, for a periodic range of the parameter q, are demonstrated. In the chosen range of q, a single desynchronous α-region of the positive TLE dominates. This is the largest hatched area. The corresponding desynchronous σ-regions, for the array of $N=10$ systems are in gray. We can see that one α-region produced five σ-regions (some of them are partly overlapped) due to independent transverse modes corresponding to five different eigenvalues γ_k (four of them are twice-degenerated). Therefore, we can observe various desynchronizing bifurcations associated with different transverse modes. Each of these σ-regions is the effect of crossing the α-region by different eigenvalues of $\mathbf{G}$, when the coupling increases. This effect is explained in detail in Fig. 4.20, where a graph of the desynchronizing mechanism analogous to the one shown in Fig. 4.17 is presented. As demonstrated, a single desynchronous interval of the MSF is precisely reflected on the bifurcation diagram of the *synchronization error* $e(\sigma)$ obtained from the direct numerical simulation of the synchronization process in the analyzed array of oscillators. However, only four separated σ-intervals are visible, because the first of them (σ^5_{1-}, σ^4_{1+}) is composed of two intervals (σ^5_{1-}, σ^5_{1+}) and (σ^4_{1-}, σ^4_{1+}), which are partly overlapped.

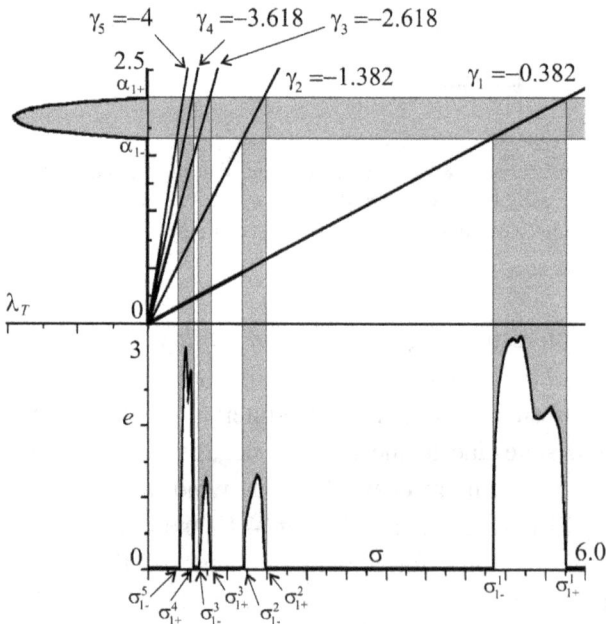

Fig. 4.20. Projection from the MSF $\lambda_T(\alpha)$ diagram, via eigenvalues γ_k, to the bifurcation diagram of the synchronization error e versus the coupling coefficient σ for an array of $N=10$ oscillators, $q=0.8$. The desynchronous intervals are in gray.

The results presented in Fig. 4.19 show that the characteristic feature of the ragged synchronizability is the self-similarity of desynchronous gray regions of the coupling parameter with a scale defined by Eq. (3.41). Thus, if the number of different eigenvalues of the connectivity matrix increases, due to the increasing number of oscillators or a varying topology of the network connections, then a cascade of self-similar desynchronous σ-regions appear and the effect of the ragged synchronizability can be observed.

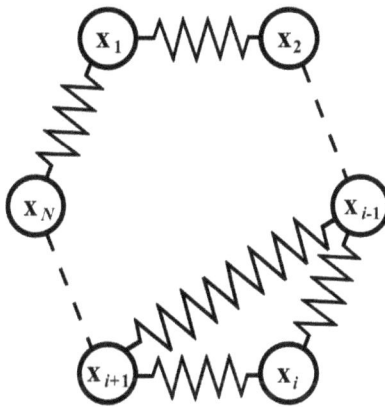

Fig. 4.21. Chain of Duffing oscillators with a shortcut.

The phenomenon of the ragged synchronizability reveals another interesting property, i.e., its sensitivity to even small changes in the topology of connections. Consider an array of ten oscillators, the same as previously (Eq. (4.40)). It is a typical regular small-world network, where we can distinguish five different eigenvalues of the connectivity matrix: -0.382, -1.382, -2.618, -3.618, -4. Let us randomize it slightly, introducing the shortest possible shortcut and avoiding only one oscillator (see Fig. 4.21), i.e., between the nodes numbered $i-1$ and $i+1$. Then, Eq. (4.40) for these nodes includes an additional component $\sigma(x_{i+1} - x_{i-1})$ or $\sigma(x_{i-1} - x_{i+1})$, respectively. This shortcut introduces four new eigenvalues: -0.504, -1.780, -3.220, -4.496. As we have shown above, in the systems exhibiting the disconnected synchronous ranges, almost each eigenvalue γ_k can influence the distribution of desynchronous σ-intervals, which can lead to considerable enhancement (or reduction) of the synchronizability. In Fig. 4.22 desynchronous σ-regions of the considered array without a shortcut are shown in black and additional regions (intervals) of desynchronization that appeared after the shortcut creation are shown in gray. These new regions are an effect of the second mode desynchronizing bifurcation in particular, which corresponds to the new eigenvalue $\gamma_2 = -0.504$ of the connectivity matrix for the array with a new link. It is clearly visible that the shortcut caused an increase in the total area of desynchronous intervals up to 30% in certain ranges of the control parameter q. Thus, slight perturbations of the network connectivity distribution can induce a significant change in its synchronizability.

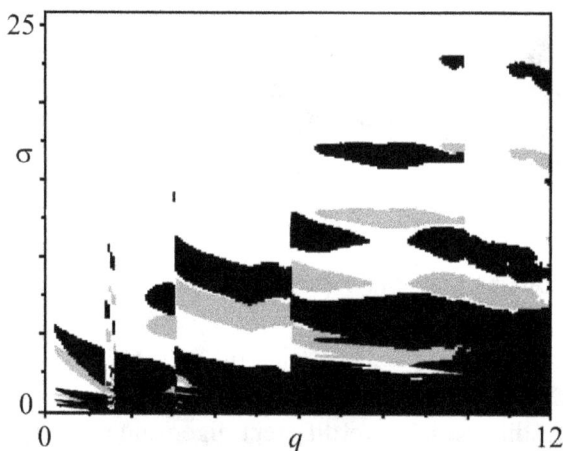

Fig. 4.22. Desynchronous σ-regions as a function of the parameter q in an array $N=10$ oscillators without a shortcut (in black) and additional regions of desynchronization which have appeared after the shortcut generation (in gray).

The total desynchronous range for any case (a chaotic or periodic node system) of the *real coupling* is defined by the following general formula:

$$\sigma \in \left\{ \left[\left(\frac{\alpha_{11}}{\gamma_2}, \frac{\alpha_{12}}{\gamma_2} \right) \cup \ldots \cup \left(\frac{\alpha_{s1}}{\gamma_2}, \frac{\alpha_{s2}}{\gamma_2} \right) \right] \cup \ldots \right.$$
$$\left. \ldots \cup \left[\left(\frac{\alpha_{11}}{\gamma_N}, \frac{\alpha_{12}}{\gamma_N} \right) \cup \ldots \cup \left(\frac{\alpha_{s1}}{\gamma_N}, \frac{\alpha_{s2}}{\gamma_N} \right) \right] \right\} . \quad (4.44)$$

For many cases of networks, desynchronous σ-ranges are completely or partly overlapped, according to formula (4.44). Sometimes this can even lead to the existence of one compact desynchronous interval $(0, \sigma_{s+}^{N-1})$, so the ragged synchronizability effect cannot be observed in this case. For instance, in the case shown in Fig. 4.17, the fulfillment of the inequalities:

$$\frac{\alpha_{2+}}{\alpha_{2-}} > \frac{\gamma_2}{\gamma_1} > \frac{\alpha_{2-}}{\alpha_{1+}}$$

causes a complete disappearance of the synchronous window (or windows) between σ^1_{1+} and σ^1_{2-} Then, only one desynchronous interval $(0, \sigma^1_{2+})$ exists.

From the above analysis, it results that a necessary (but not sufficient) condition for the ragged synchronizability is the existence of at least one double-limited MSF-interval of the positive TLE (e.g., see Fig. 4.20) with non-zero boundaries (i.e. α_{j-}, $\alpha_{j+} > 0$). The source of this interesting effect is a folding or bubbling character of the MSF (Figs. 4.16a – 4.16c). Such a form of the MSF results in a cascade of self-similar desynchronous intervals of coupling strength (Figs. 4.19 and 4.20). Between them, synchronous windows are located, which is the essence of the ragged synchronizability. Its occurrence is independent of the motion character (periodic or chaotic) of the isolated node system. Morcover, a rich spectrum of desynchronizing bifurcations corresponding to different transverse modes and sensitivity of small changes in a topology of network links (Fig. 4.22) are characteristic features that accompany this phenomenon.

4.2.2 *Experimental observation of disconnected synchronous ranges. Influence of a parameter mismatch*

In the experimental studies supported by the numerical analysis (Perlikowski (2007), Perlikowski *et al.* (2008)), the VdP oscillator:

$$\dot{x} = y,$$
$$\dot{y} = x - h(1 - x^2)y + q\cos(\eta t), \tag{4.45}$$

where h, q and η are constant, has been taken as an array node. The constants q and η represent the amplitude and the frequency of the external excitation, respectively. Consider an open array of three mutually coupled VdP oscillators schematically illustrated in Fig. 4.23. The corresponding connectivity matrix is:

$$\mathbf{G} = \begin{pmatrix} -2 & 2 & 0 \\ 2 & -3 & 1 \\ 0 & 1 & -1 \end{pmatrix}, \tag{4.46}$$

and the ND coupling between the oscillators is realized according to the output function given by Eq. (4.41). Hence, the evolution of all oscillators coupled in this array is given by the following set of differential equations:

$$\dot{x}_1 = y_1,$$
$$\dot{y}_1 = x_1 - h(1 - x_1^2)y_1 + q\cos(\eta t) + 2\sigma(x_2 - x_1),$$
$$\dot{x}_2 = y_2,$$
$$\dot{y}_2 = x_2 - h(1 - x_2^2)y_2 + q\cos(\eta t) + \sigma(2x_1 + x_3 - 3x_2), \tag{4.47}$$
$$\dot{x}_3 = y_3,$$
$$\dot{y}_3 = x_3 - h(1 - x_3^2)y_3 + q\cos(\eta t) + \sigma(x_2 - x_3),$$

where σ is a constant coupling coefficient. Substituting the analyzed system Eq. (4.45) into Eq. (3.40), we obtain a generic variational equation for computing the MSF in the form:

$$\dot{\xi} = \psi,$$
$$\dot{\psi} = h(1 - x^2)\psi - 2hx\xi y - \xi + \sigma\gamma\xi. \tag{4.48}$$

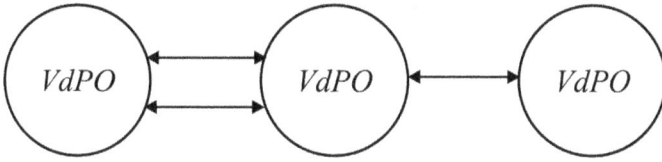

Fig. 4.23. Model of an open array of van der Pol's oscillators (VdPO).

In the numerical analysis, we assume $h=0.401$, $\eta = 1.207$, and consider σ as a control parameter. Then, in the absence of a coupling, each oscillator exhibits periodic behavior ($\lambda_1 = -0.126$) with the period equal to the period of excitation.

In the experiment, we have used an electronic implementation of this array with real equivalents of dimensionless parameters, which is schematically depicted in Fig. 4.2. Each VdP oscillator has been implemented as a circuit (Nana & Woafo (2006), shown in the black frame in Fig. 4.24) composed of two capacitors $C1$ and $C2$, seven resistors $R1-R7$, and two multiplicators AD-633JN which introduce nonlinearity. Multiplicators have the following characteristics: $W= (1/V_c)(X_1 - X_2)(Y_1 - Y_2) + Z$, where X_1, X_2, Y_1 and Y_2 are input signals, W is an output signal, and $V_c = 10$ [V] is a characteristic voltage. The input $E_m \cos\omega t$, where the amplitude E_m and the frequency ω are constant, represents external excitation. The additional resistors $R8$ and R have been used to realize the coupling. In our implementation we have used "out-of-shelf" elements: $R1=9920$ [Ω], $R2=999$ [Ω], $R3=501$ [Ω], $R4= 100$ [Ω], $R5=10^4$ [Ω], $R6=10^4$ [Ω], $R7=16.15\times10^4$ [Ω], $R=18\times10^4$ [Ω], $C1=10$ [nF], $C2=10$ [nF]. $R8 \in (0$ [Ω], 44×10^4 [Ω]) has been taken as a control parameter. The equivalent elements in each circuit can differ by 1% of their nominal values. The relation between the circuit real parameters and the dimensionless parameters of Eqs. (4.47) is as follows:

$$\omega_0^2 = \frac{1}{C1C2R2R7}, \quad h = \frac{1}{C1R1\omega_0^2}, \quad \eta = \frac{\omega}{\omega_0},$$

$$x_{1-3} = V_{1-3}\frac{R3}{E_m R2}, \quad y_{1-3} = \delta V_{1-3}\frac{R3}{E_m R2\omega_0}, \quad \sigma = \frac{R2}{R8}.$$

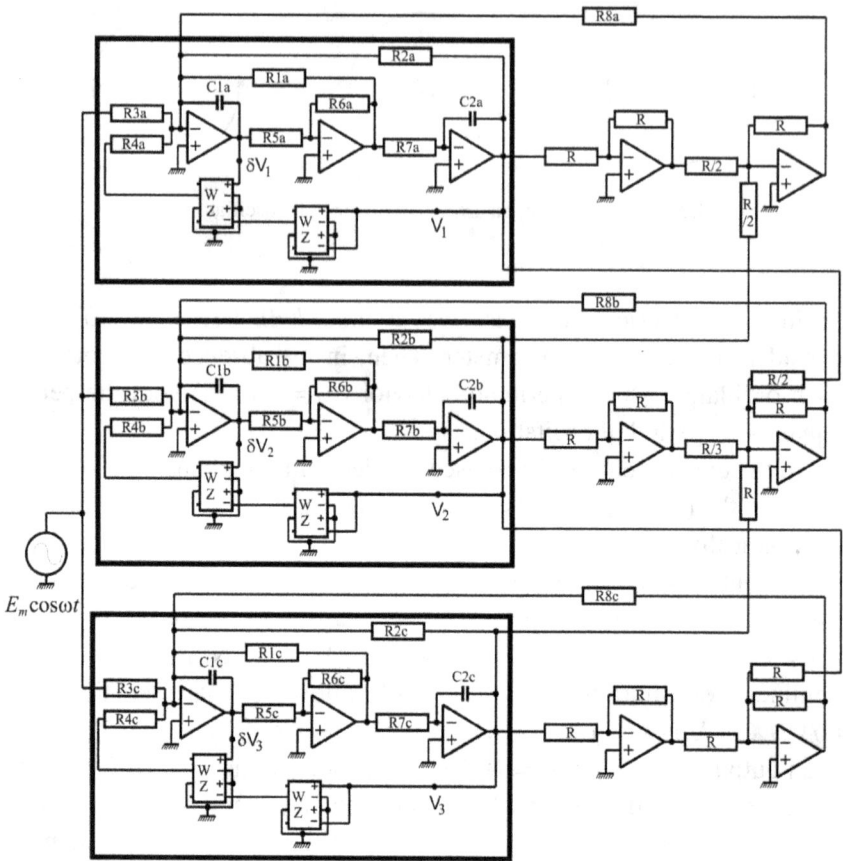

Fig. 4.24. Electronic implementation of an open array of the VdP oscillators shown in Fig. 4.23.

The nonidentity of elements used in each circuit introduces the mismatches of d and σ parameters in Eqs. (4.47). The estimated mismatches are smaller than ± 0.001. The data are acquired with a Data Acquisition System 3200A\415 board connected to a computer controlled by the software developed in Microstar Labs. The dynamical variables of interest in this circuit are the voltages V_{1-3} of each oscillator measured in the points indicated in Fig. 4.24. The first derivatives of the potentials V_{1-3} are taken in the point also indicated in Fig. 4.24.

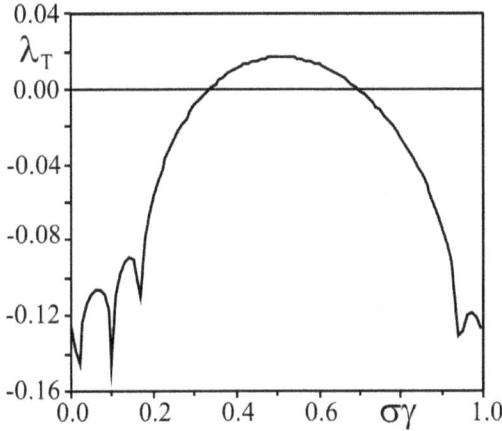

Fig. 4.25. Largest TLE λ_T, calculated for the generic variational equation (Eq. (4.48)) versus the product $\sigma\gamma$, η=1.207, h=0.401.

The matrix **G** has three real eigenvalues $\gamma_0 = 0$, $\gamma_1 = -1.27$, $\gamma_2 = -4.73$, so this is a variant of the *diffusive real coupling*. The MSF graph for the presented example, the largest TLE versus product $\sigma\gamma$ (i.e., α) is depicted in Fig. 3. If the products $\sigma\gamma_{1,2}$ corresponding to both transversal eigenmodes can be found in the ranges of the negative transversal Lyapunov exponent, then the synchronous state is stable for the analyzed configuration of couplings. Looking at this MSF diagram in Fig. 4.25, we can expect an appearance of the ragged synchronizability of coupled VdP oscillators (Eq. (4.45)), because two disconnected regions of the negative TLE, i.e., $\sigma\gamma \in ((0, \ 1+) \ \cup \ (0, \ 1-))$, can be observed. Consequently, at least two separated synchronous ranges of the coupling parameter σ should be visible, i.e., the ragged synchronizability effect takes place. Its mechanism in the case under consideration is graphically explained in Figs. 4.26a–c in a way known from the previous Sec. 4.2.1. Additionally, a comparison of the numerical (including the parameter mismatch, Fig. 4.26c) and experimental (Fig. 4.26d) synchronization is demonstrated. In both the analyzed cases (Figs. 4.26c and 4.26d), the synchronization error is defined as follows:

$$e = \sum_{i=2}^{3} \left| x_1 - x_i \right|. \tag{4.49}$$

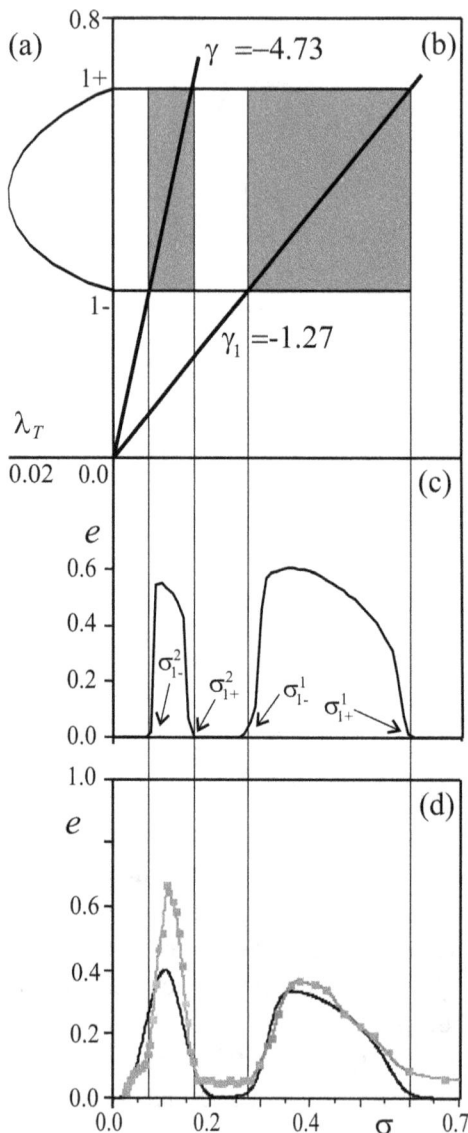

Fig. 4.26. Comparison of the numerical and experimental results, $h = 0.401$, $\eta = 1.207$; (a)–(c) desynchronizing mechanism: A projection from the MSF diagram (a), via eigenvalues γ_k of the connectivity matrix $\mathbf{G}$ (b), to the bifurcation diagram of synchronization error e versus the coupling coefficient σ (c) (desynchronous intervals are shown in gray), (d) synchronization error e versus σ; the experimental results — a gray line with scatters (marking measurement points) and the numerical results for the case of the parameter mismatch (black line).

We can observe the third synchronous $(\sigma^2_{1+}, \sigma^1_{1-})$ and the second desynchronous $(\sigma^2_{1-}, \sigma^2_{1+})$ σ intervals in comparison with only two synchronous and one desynchronous MSF ranges, respectively. An additional desynchronous interval appears because mode 2 (associated with the eigenvalue γ_2) crosses the desynchronous MSF interval (1−, 1+), while mode 1 (associated with γ_1) is still located in the first synchronous MSF interval (0, 1−), (see Fig. 4.26b). Then, in the narrow range $(\sigma^2_{1+}, \sigma^1_{1-})$, two modes are in the synchronous MSF interval, so that one can observe an "additional window of synchronization" in the σ interval. The second desynchronous σ interval corresponds to the mode 1 desynchronizing bifurcation. Finally, the steady synchronous state is achieved due to the increasing coupling strength at $\sigma=0.6$.

In Fig. 4.26d results of the experimental investigation of the synchronization process in the analyzed circuit are demonstrated. The plot of the experimentally generated *synchronization error*, reduced to a nondimensional form and calculated with the use of Eq. (4.49) versus the coupling strength, is shown in gray with scatters. Obviously, in the case of real VdP oscillators, the perfect CS cannot be achieved due to the unavoidable parameter mismatch. However, in such a case the ICS (Eq. (1.4)) can be observed, when the correlation of amplitudes and phases of the system responses is not ideal, but a synchronization error remains relatively small during the time evolution. One can notice a qualitative coincidence between the numerical simulations and the experiment comparing Fig. 4.26c to Fig. 4.26d, i.e., the regions of the ICS tendency in a real circuit correspond well to the complete synchronization ranges in the numerical model. In the last stage of our research, an the influence of the parameter mismatch on the synchronization error e has been analyzed numerically. We have estimated a slight disparity of the values of h in all three VdP oscillators while measuring their real parameters. Next, such an approximated mismatch has been realized in the considered system (4.47) (the values of all three h taken in Eqs. (4.47) are, respectively, 0.400, 0.401, and 0.402). The synchronization error simulated numerically for this model is represented with a black line in Fig. 4.26d. Its good visible agreement with the experimental result shows that a slight difference of coupled oscillators does not destroy their synchronization tendency, i.e., the ICS takes place. Comparing the analytical (obtained by the MSF approach, Fig. 4.26b), numerical (Fig. 4.26c) and experimental results (Fig. 4.26d), we can confirm an occurrence of the ragged synchronizability phenomenon in the real system of coupled oscillators. It seems that the phenomenon of RSA is common for the systems with a *non-diagonal coupling* and not sensitive to a small parameter mismatch, i.e., it can be observed in real experimental systems.

4.2.3 *Designing the ragged synchronizability*

The above-demonstrated numerical and experimental outcomes have shown that the ragged synchronizability is a property of the networks rather with the ND than the CD or PD coupling between nodes. However, an accomplishment of this interesting effect is also possible when only the PD coupling is applied to connect the oscillators. In order to exemplify such a case, let us consider Lorenz oscillators again:

$$\dot{x} = -\delta x + \delta y,$$
$$\dot{y} = -xz + rx - y, \tag{4.50}$$
$$\dot{z} = xy - bz,$$

analogous to those used in double system (4.38). Now, introduce a non-uniform (i.e., diagonal components are different) PD xz-coupling between them, defined by the following output function:

$$\mathbf{H} = \begin{pmatrix} 1 & 0 & 0 \\ 0 & 0 & 0 \\ 0 & 0 & 4 \end{pmatrix}. \tag{4.51}$$

The variational equation for calculating the MSF (Eq. (3.40)) is as follows:

$$\dot{\xi} = -\delta\xi + \delta\psi - \alpha\xi,$$
$$\dot{\psi} = -x\zeta + (r - z)\xi - \psi, \tag{4.52}$$
$$\dot{\zeta} = \xi y - b\zeta - 4\alpha\zeta.$$

The MSF computed from Eqs. (4.50), (4.52) and the corresponding bifurcation diagram of the double Lorenz system (4.38) are illustrated in

Figs. 4.27a and 4.27b, respectively. The graph of the *trajectory separation* x_1-x_2 versus the coupling coefficient σ shown in Fig. 4.27b has been reconstructed from system (4.38) according to linking function (4.51), i.e., $d_x = \sigma$, $d_y = 0$, $d_z = 4\sigma$. Here, the existence of the *ragged synchronizability* is clearly visible, i.e., two synchronous ranges of the negative TLE, $(\alpha_{1+}, \alpha_{2-})$ and $(\alpha_{1+}, \alpha_{2-})$, coexist with two desynchronous intervals of the positive TLE, $(0, \alpha_{1+})$ and $(\alpha_{2-}, \alpha_{2+})$. This situation is precisely reflected in bifrcation diagram 4.27b, where x_1-x_2 approaches zero for $\lambda_T < 0$.

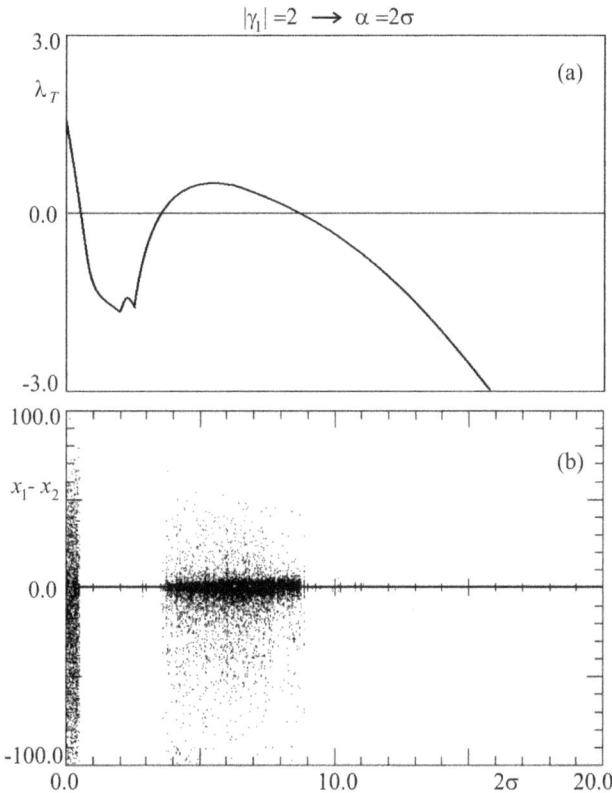

Fig. 4.27. MSF of the Lorenz system (4.50) with the PD linking function (4.51) computed from variational equation (4.52) (a) and the corresponding bifurcation diagram of the *trajectory separation* x_1-x_2 versus the coupling coefficient σ for the double Lorenz system (4.38) (b). Since $|\gamma_1| = 2$ for two mutually coupled oscillators, thus we have $\alpha = 2\sigma$.

The idea of such designing the *ragged synchronizability* consists in a combination of two PD coupling schemes (with a single diagonal component) in such a way that one of them leads to the bottom-limited synchronizability, while the second one causes a double (upper) bounded synchronous interval. A comparison of the MSFs of the couplings used in the above example is presented in Fig. 4.28. We can see that the *x*-coupling of Lorenz systems (4.50) with the finally negative TLE $\lambda_T(x)$ (in gray) has been combined with a four-times stronger *z*-coupling characterized by the positive TLE $\lambda_T(4z)$ for the larger coupling strength (in black). The MSF shown in Fig. 4.27a seems to be an effect of the common interaction of *x* and *4z* couplings. After the first interval of the positive TLE $(0, \alpha_{1+})$ visible in diagram 4.27a, there appears its negative region $(\alpha_{1+}, \alpha_{2-})$ as a reflection of the negative $\lambda_T(4z)$ interval in Fig. 4.28. But further on, even a slight increase in α causes a significant growth of $\lambda_T(4z)$, while $\lambda_T(x)$ still remains relatively large and, as a result, the second desynchronous interval of the MSF appears (Fig. 4.27a). Finally, the increased α stabilizes the synchronization in the range $\alpha > \alpha_{2+}$. Obviously, such a common interaction of the PD coupling effects is not a "linear" superposition of them, but, as we could see, it can be also a reason of the *ragged synchronizability*.

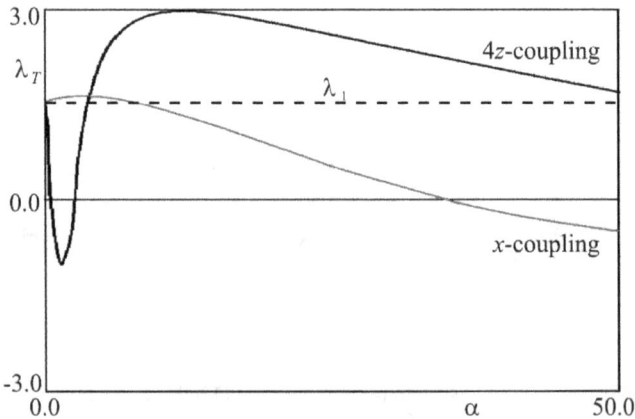

Fig. 4.28. Comparison of the *x*-coupling MSF (in gray) with the *4z*-coupling MSF (in black) for Lorenz systems (4.50).

4.3 Commonly Driven Oscillators

In this Sec., the results of numerical investigations of the CS and GS in the systems coupled according to the scheme shown in Fig. 2.4 and described by Eqs. (2.22) and (2.23) are demonstrated. The idea of the GS is applied here rather as a tool for analysis of the cooperative motion of driven (slave) oscillators, which manifests their CS. Generally, the GS problems have been researched both in the context of identical master system and slave subsystems (when separated), and also in cases when the response system is slightly (the same set of ODEs with different values of system parameters) or strictly different (another set of ODEs) than the driving oscillator (Abarbanel *et al.* (1996), Kocarev & Parlitz (1996), Pyragas (1996), Afraimovich *et al.* (2002), Boccaletti *et al.* (2002), Femat *et al.* (2005), Stefanski (2008)). These both types of the coupling via a common signal are considered below.

4.3.1 *Master-slave coupling of identical systems*

In order to demonstrate the synchronization properties of commonly driven identical oscillators, two examples of standard chaotic systems in a master-slave configuration of the coupling have been considered. They are diffusively and unidirectionally coupled discrete-systems or continuous-time systems, described in the general form as follows:

$$\mathbf{x}_{n+1} = \mathbf{f}(\mathbf{x}_n), \tag{4.53a}$$

$$\mathbf{y}_{n+1} = \mathbf{f}(\mathbf{y}_n) + d[\mathbf{f}(\mathbf{x}_n) - \mathbf{f}(\mathbf{y}_n)], \tag{4.53b}$$

$$\mathbf{y}'_{n+1} = \mathbf{f}(\mathbf{y}'_n) + d[\mathbf{f}(\mathbf{x}_n) - \mathbf{f}(\mathbf{y}'_n)], \tag{4.53c}$$

or

$$\dot{\mathbf{x}} = \mathbf{f}(\mathbf{x}), \tag{4.54a}$$

$$\dot{\mathbf{y}} = \mathbf{f}(\mathbf{y}) + d(\mathbf{x} - \mathbf{y}), \tag{4.54b}$$

$$\dot{\mathbf{y}}' = \mathbf{f}(\mathbf{y}) + d(\mathbf{x} - \mathbf{y}), \tag{4.54c}$$

where $\mathbf{y}'$ represents an auxiliary response system for detecting the GS between the drive and response oscillators. As has been mentioned in

Sec. 1.3, the GS can be recognized in its strong and weak version, which in the general case can be defined by equality (1.9) and inequality (1.8), respectively. However, for a particular case of completely identical drive-response systems, the criteria for the GS are simpler, because it can be detected in direct simulations of the synchronization process in systems (4.53a–c) or (4.54a–c). Namely, the strong GS means the CS between the master system (4.53a) or (4.54a) and response oscillators (4.53b, c) or (4.54b, c). On the other hand, the weak GS occurs when the CS of the response **y** and its auxiliary replica **y'** can be observed, whereas their synchronized motion remains uncorrelated with the drive **x**.

As the first example, let us consider one-dimensional logistic maps introduced into Eqs. (4.53a–c):

$$x_{n+1} = rx_n(1-x_n), \qquad (4.55a)$$

$$y_{n+1} = (1-d)[ry_n(1-y_n)] + d[rx_n(1-x_n)], \qquad (4.55b)$$

$$y'_{n+1} = (1-d)[ry'_n(1-y'_n)] + d[rx_n(1-x_n)]. \qquad (4.55c)$$

The numerical analysis of various synchronization aspects in system (4.55a–c) is graphically depicted in Figs. 4.29a–d, where bifurcation diagrams of the *synchronization error* (Figs. 4.29a–c) and Lyapunov exponents (Fig. 4.29d) versus the coupling parameter are presented. Here, gray and black dashed lines mark the thresholds the weak and strong GS, respectively. Comparing the diagrams from Figs. 4.29a and 4.29b, we can see that the weak GS ($y - y'$) appears below the threshold of the strong one ($x - y$). Such a weakness of the GS is confirmed by the bifurcation diagram shown in Fig. 4.29c, which has been evaluated for marginally different response oscillators, i.e., the mismatch of parameter r between the response systems y and y' amounts $\Delta r = 10^{-12}$. However, even such an "invisible" parameter disturbance causes a destruction of the weak synchrony regime. Then, the synchronization threshold is determined by the ICS between y and y'. The comparison of Figs. 4.29a and 4.29c demonstrates that this ICS limit parameter value approximately coincides with the strong GS threshold. Thus, it reflects the properties of weak and strong versions of the GS. On the other hand, a quantitative criterion for both cases of the GS is given by the spectrum of three Lyapunov exponents of triple system (4.55a–c) shown in Fig. 4.29d. The first of them λ_1 (dashed line) is the LLE of the driving oscillator (4.55a), which is obviously independent of the coupling parameter. Therefore, it has a constant value in spite of the increasing coupling strength. The second one (gray curve) is the TLE of the identity manifold $x=y$:

$$\lambda_T = \ln(1-d) + \lim_{n \to \infty} \frac{1}{n} \sum_{i=1}^{n} \ln|f(x_i)|, \qquad (4.56)$$

which in the case under consideration is reduced to the form:

$$\lambda_T = \ln(1-d) + \lambda_1. \qquad (4.57)$$

According to the known theory (Pyragas (1998)), a negativity of λ_T is the condition for the strong GS. The irregular black line in Fig. 4.29d represents the RLE (λ_R) of the drive–response system (4.55a–b), which is defined by the formula:

$$\lambda_T = \ln(1-d) + \lim_{n \to \infty} \frac{1}{n} \sum_{i=1}^{n} \ln|f(y'_i)|. \qquad (4.58)$$

From the comparison of Figs. 4.29b and 4.29d, it results that the weak GS in system (4.55a–c) takes place if the largest RLE (λ^1_R) is negative in the coupling parameter space.

The second example are x-coupled, chaotic Lorenz oscillators (Eqs. (4.38) or (4.50) — analyzed above) introduced into Eqs. (4.54a–c):

$$\dot{x}_1 = -\delta x_1 + \delta x_2,$$
$$\dot{x}_2 = -x_1 x_3 + rx_1 - x_2, \qquad (4.59a)$$
$$\dot{x}_3 = x_1 x_2 - bx_3,$$

$$\dot{y}_1 = -\delta y_1 + \delta y_2 + d(x_1 - y_1),$$
$$\dot{y}_2 = -y_1 y_3 + ry_1 - y_2, \qquad (4.59b)$$
$$\dot{y}_3 = y_1 y_2 - by_3,$$

$$\dot{y}'_1 = -\delta y'_1 + \delta y'_2 + d(x_1 - y'_1),$$
$$\dot{y}'_2 = -y'_1 y'_3 + ry'_1 - y'_2, \qquad (4.59c)$$
$$\dot{y}'_3 = y'_1 y'_2 - by'_3.$$

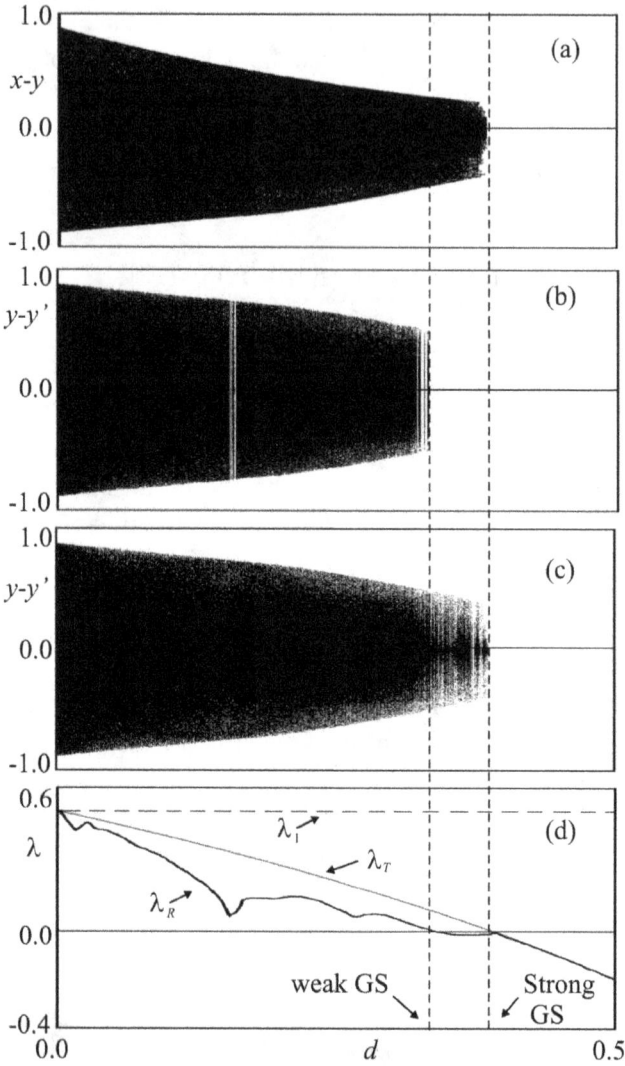

Fig. 4.29. Bifurcation diagrams of the synchronization error versus the coupling coefficient d in the system of unidirectionally coupled logistic maps (Eqs. (4.55a–c)). Identical drive response oscillators x–y (a), identical response oscillators y–y' (b), slightly different response oscillators y–y'– $\Delta r=10^{-12}$ (c). Corresponding Lyapunov exponents of system (4.55a–c): the DLE (λ_1) — dashed line, the TLE (λ_T) — gray curve, the RLE (λ_R) — black line. Parameter $r = 3.9$.

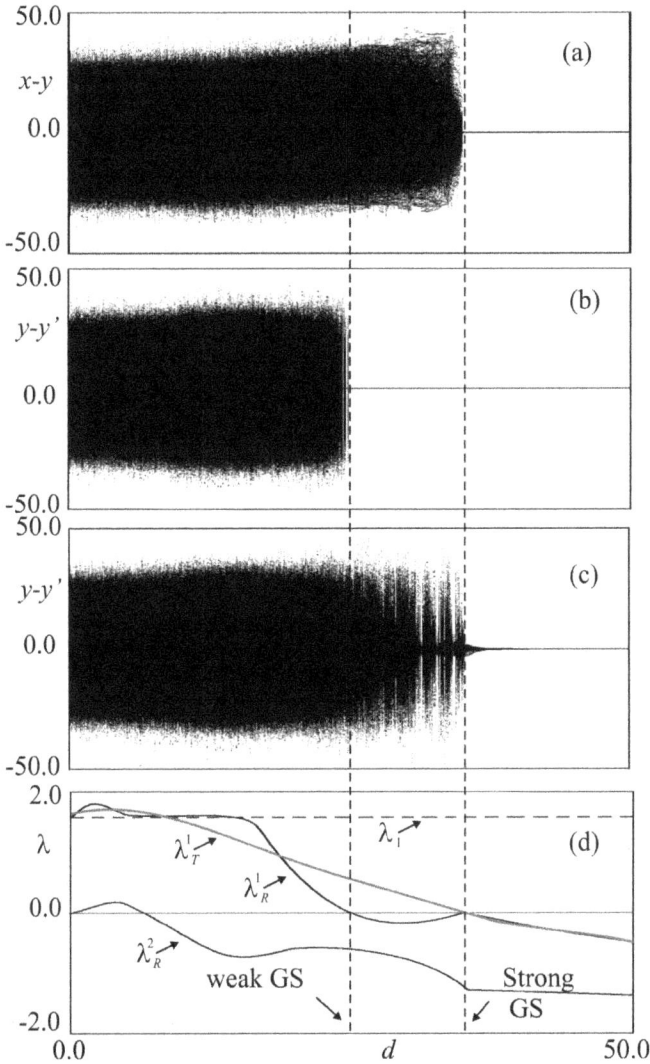

Fig. 4.30. Bifurcation diagrams of the synchronization error versus the coupling coefficient d in the system of unidirectionally coupled Lorenz oscillators (Eqs. (4.59a–c)). Identical drive response systems x–y (a), identical response oscillators y–y' (b), slightly different response oscillators y–y' — $\Delta r = 10^{-4}$ (c). Corresponding Lyapunov exponents of system (4.59a–c): the DLE (λ_1) — dashed line, and two the largest RLE (λ^1_R, λ^2_R) — black line. System parameters: $\delta = 10.0$, $b = 8/3$, $r = 120.0$.

The results of the numerical analysis of the synchronous behavior in system (4.59a–c) are demonstrated in Figs. 4.30a–d. We can observe a full analogy to the case of discrete-time systems (4.55a–c). The thresholds of the weak (Fig 4.30a) and strong (Fig 4.30b) synchronization are determined by the negativity of the largest RLE and TLE (Fig 4.30d), respectively. Moreover, an introduction of the small parameter mismatch, i.e., a perturbance of the control parameter r in the response Lorenz systems y and y' is $\Delta r = 10^{-4}$, also causes a desychronous effect (Fig. 4.30c), similarly to the logistic map case (Fig. 4.29c). Thus, it is a proof that the weak GS regime is very sensitive to the mismatch of response oscillators (Pyragas (1998)). These results show that a real condition for the strong GS of identical drive–response systems is $\lambda^1_R < 0$, in spite of the fact that the dimension of the global attractor is still larger than the dimension of the driving system attractor, i.e., the strong GS condition (Eq. (1.9)) is not fulfilled. For instance, the dimension of the driving attractor (4.59a) is $d^D = 2.101$, while the global dimension of the drive–response system (4.59a, b) is $d^G > 4$ even in the range $\lambda^1_T > 0$, where the strong GS takes place (see Figs. 4.30a and 4.30c). Consequently, the GS condition (1.9) does not operate in each drive–response configuration.

4.3.2 Non-identical drive–response systems

In this Sec., the numerical investigations of the CS in arrays of identical response oscillators driven by the same external signal, which are described in the general form by Eqs. (2.22) and (2.23), are presented. A case of a rather strict difference between the master oscillator (see Eqs. (3.59a) or (3.60a)) and the slave sub-system is studied. Then, the dynamics of each i-th ($i=1, 2,\ldots, N$) individual master–slave component in the array is given by:

$$\dot{\mathbf{e}} = \mathbf{g}(\mathbf{e}), \tag{4.60a}$$

$$\dot{\mathbf{x}}_i = \mathbf{f}(\mathbf{x}_i) + q\mathbf{h}(\mathbf{e}), \tag{4.60b}$$

for flows, and

$$\mathbf{e}_n = \mathbf{g}(\mathbf{e}_n), \qquad (4.61a)$$

$$\mathbf{x}_{n+1}^i = \mathbf{f}\left(\mathbf{x}_n^i\right) + q\mathbf{h}(\mathbf{e}_n), \qquad (4.61b)$$

for maps, respectively. In the bifurcation analysis, the coefficient q representing the amplitude of the drive, has been used as a control parameter.

The numerical experiment has been carried out by means of the DYNAMICS (Nusse & Yorke (1997)) and DELPHI[1] software. Since the problem of periodically forced oscillators has been widely described (Van der Pol (1927), Timoshenko (1928), Den Hartog (1934) Blekhman (1988)), the analysis of the CS is concentrated on oscillators with a chaotic and stochastic external drive. In the numerical simulations reported below, non-linear mechanical oscillators of the Duffing type have been used as the response system, which is forced by an irregular *deterministic* drive (taken from the chaotic Lorenz system) or random excitation, generated from the Gaussian process, working as a *non-deterministic* drive. Furthermore, a discrete-time response is analyzed: the chaotic Henon map driven by the chaotic logistic map or the random function. This analysis is a confirmation of the theoretical considerations from Sec. 3.3.3. On the other hand, it is also a verification and exemplification of the proposed method of the largest RLE estimation, presented in detail in Part II of this book (Secs. 5–7). Therefore, all demonstrated examples of the application of this synchronization-based method are illustrated with parameters of its numerical procedure.

4.3.2.1 *Array of driven mechanical oscillators*

The system under consideration is an array of N identical non-linear oscillators of the Duffing type with a kinematic drive, as shown in Fig. 4.31. The dynamics of a single i-th oscillator (Eq. (4.60b)) is governed by the following dimensionless equations of motion:

$$\dot{x}_{i1} = x_{i2},$$
$$\dot{x}_{i2} = -\alpha x_{i1}^3 - h x_{i2} + q[e(t) - x_{i1}], \qquad (4.62)$$

---

[1] www.borland.com

where the parameters α, q and h are dimensionless representations of the real parameters c, c_1 and μ (see Fig. 4.31), respectively, and $e(t)$ is an arbitrary driving signal taken from Eq. (4.60a). In all the numerical experiments presented here, values $\alpha = 10.0$ and $h = 0.3$ were taken.

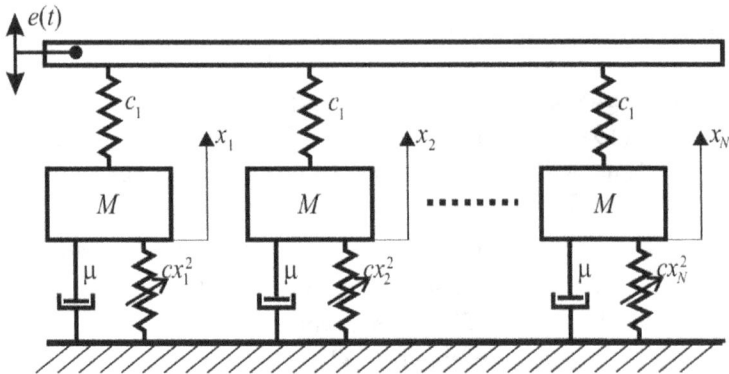

Fig. 4.31. Mechanical oscillators of the Duffing type with a kinematic external driving; M — mass of the oscillators, μ — coefficient of viscous damping, c — nonlinear spring rate, c_1 — stiffness of the linear spring transmitting a driving signal.

The character of the problem under investigation results in some qualitative differences in comparison with other researchers' works. Namely, in the literature dealing with the GS, the approaches where the *master–slave coupling* of chaotic systems (when uncoupled) is analyzed, dominate (Abarbanel *et al.* (1996), Kocarev & Parlitz (1996), Pyragas (1996), Afraimovich *et al.* (2002), Boccaletti *et al.* (2002), Femat *et al.* (2005), Stefanski (2008)). Usually these systems are autonomous chaotic Rössler and Lorenz oscillators (mentioned here many times) or self-excited electrical circuits (Van der Pol (1920), Chua oscillators (1986)). Such systems possess their own energy source and can oscillate independently of an additional external drive, i.e., motion of the response oscillator takes place also without a coupling with the driving system. On the other hand, a "pure external drive" case is considered here, i.e., the driving system (Eq. (4.60a)) is the only one source of energy for the response system. Hence, if this system is isolated from the drive ($q=0$), then the solution to Eq. (4.62), representing response system (4.60b), tends to a stable fixed point in the phase space due to the presence of damping signified with the coefficient h in Eq. (4.62). Then, a typical case of free damped vibrations can be observed.

In the first numerical example, a source of external excitation (Eq. (4.60a)) is the well-known Lorenz system:

$$\dot{e}_1 = -10(e_1 - e_2),$$
$$\dot{e}_2 = -e_1 e_3 + 28 e_1 - e_2, \qquad (4.63)$$
$$\dot{e}_3 = e_1 e_2 - \frac{8}{3} e_3.$$

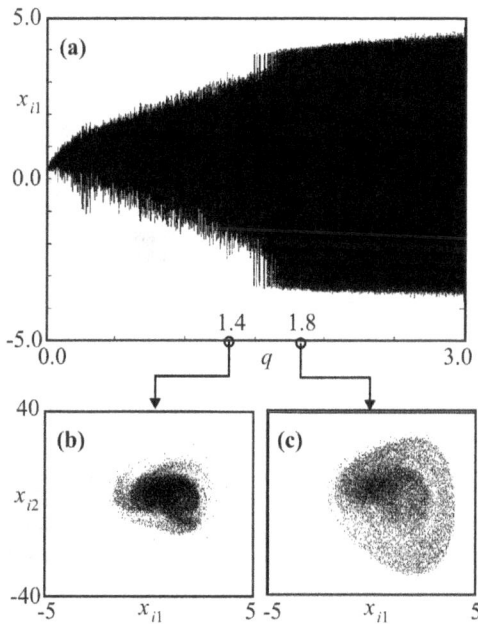

Fig. 4.32. Bifurcation diagram of the single drive response system (Eqs. (4.62) and (4.63)) (a) and exemplary Poincare cross-sections of chaotic synchronous (b) and hyperchaotic desynchronous (c) attractors.

Oscillators (4.62) are excited using a signal given by a variable e_3 from Eq. (4.63), i.e., $e(t) = e_3$. For the assumed set of parameters, the Lorenz system works in the chaotic range (the largest DLE is positive — $\lambda^1_D = 0.899$). Thus, this is a case of the chaotic *deterministic drive*.

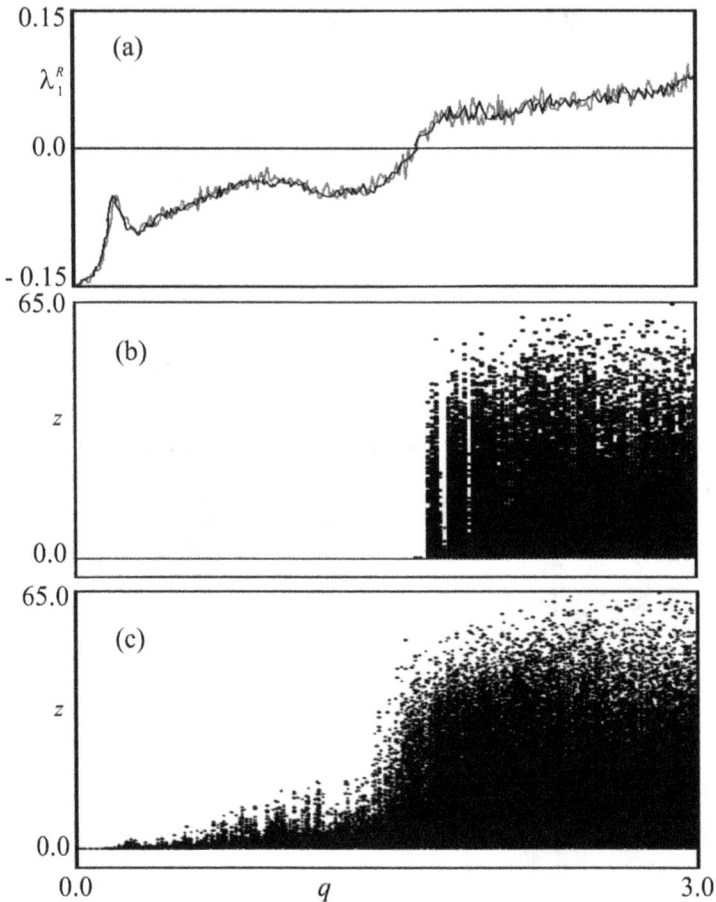

Fig. 4.33. Bifurcation diagram of the largest RLE of system (4.62) with a *deterministic drive* (Eq. (4.63)) calculated by means of the classical algorithm (in black) and estimated using the synchronization method (in gray) (a); the corresponding *synchronization error* computed according to Eq. (4.43) for $N=2$ identical (b) and slightly different (1% of a mismatch in the parameter α) response oscillators (c); Parameters of the synchronization method: $d_{1(0)} = -0.2$, $d_{2(0)} = 0.2$, $\delta = 0.001$, $\varepsilon = 0.1$, $D = 2.0$, $T_e = 10000$, $T_t = 0.5T_e$.

The results of its bifurcation analysis are presented in Figs. 4.32 and 4.33. A chaotic character of excitation causes an irregular motion of each response oscillator in the entire range of the bifurcation coefficient (Fig. 4.32a). This irregularity is also observable on the Poincare maps (Poincaré (1913)), see Figs. 4.32b and 4.32c, accompanying the bifurcation diagram, which have been generated for some chosen values of q. Figure 4.33a illustrates how the corresponding largest RLEs, calculated by means of the classical algorithm (Shimada & Nagashima (1979), Benettin *et al.* (1980a) and (1980b)) and the ones estimated with the synchronization method, vary. In order to examine the synchronizability of slave oscillators in direct simulations, it is enough to investigate the *synchronization error* (i.e., *trajectory separation*) of only two response neighbors (4.62). Such a comparison (the bifurcation diagram shown in Fig. 4.33b) with the corresponding RLEs shows that the CS appears (with decreasing q, the *synchronization error* approaches zero) when the largest RLE becomes negative, although the motion of the system is still chaotic. Obviously, the CS of the slave oscillators indicates the GS between the drive and the response. Thus, the GS in externally driven mechanical oscillators is an effect of the disappearing sensitivity of the system response to the initial condition. Then, the functional relationship (see Eq. (1.6)) defining the GS appears. In other words, if the GS takes place, then the response of the system has a regular nature (negative RLEs − a stable *sub-attractor* in the **R** subspace, see Fig. 3.9a) in spite of the observed chaotic behavior of the system, which is caused by the chaotic drive only.

Another aspect of the considered case of the synchronization phenomenon is a process of transition between chaos and hyperchaos. A loss of the CS shown in Fig. 4.33b notifies the transition from chaotic motion of the drive–response system (Eqs. (4.62) and (4.63)) on the synchronous attractor (see Fig. 4.32b) with one positive Lyapunov exponent ($\lambda^1_D > 0$, $\lambda^1_R < 0$) to hyperchaotic behavior of the system characterized by two positive Lyapunov exponents ($\lambda^1_D > 0$, $\lambda^1_R < 0$), which takes place on the desynchronous attractor (Fig. 4.32c). Such transition is manifested by a sharp increase of the attractor volume in the phase space (Figs. 4.32a–c). The next feature of the desynchronizing mechanism is the well-known phenomenon of on-off intermittency (Platt *et al.* (1993)), which has also been observed in the system under consideration.

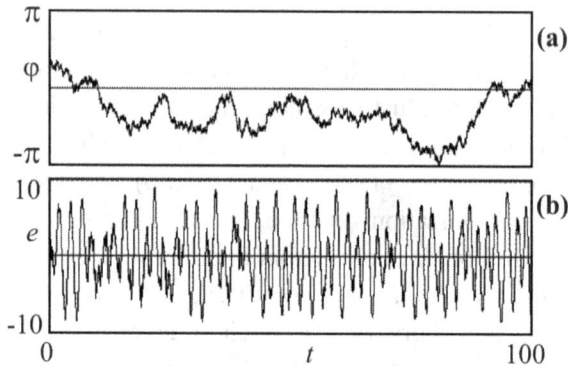

Fig. 4.34. Time diagrams of the random phase (Eq. (4.64a)) and the corresponding stochastic driving signal (Eq. (4.64b)).

From the viewpoint of practical application, it is important to know an influence of a small parameter mismatch (between the response oscillators) on the synchronization process, because in real systems their ideal identity is unlike. It is obvious that even the smallest disturbance of the system identity makes the CS impossible. However, in such a case an occurrence of the ICS (Eq. (1.4)) can be observed. In Fig. 4.33c a bifurcation diagram of the *synchronization error* for two slightly different response oscillators (1% of difference in the parameter α) is shown. Unfortunately, we can see that such a small parameter mismatch causes a significant perturbation of the synchronization process. The ICS occurs only for smaller values of the driving parameter ($q < 0.4$). It is a meaningful shift of the synchronization threshold in comparison with the ideal case of the CS shown in Fig. 4.33b.

In the second numerical example, a *non-deterministic drive* is applied to the oscillators (4.62). Such a driving signal can be generated using numerical techniques to model any stochastic process, e.g., the Gaussian process with the spectral representation method (Shinozuka (1992)). In the simulations carried out, a random number generator embedded in the

DELPHI environment has been applied. The signal has a form of the series

$$e(t) = \sum_{K=1}^{L} K \sin[K\Omega t + \gamma_K \varphi(t)],$$

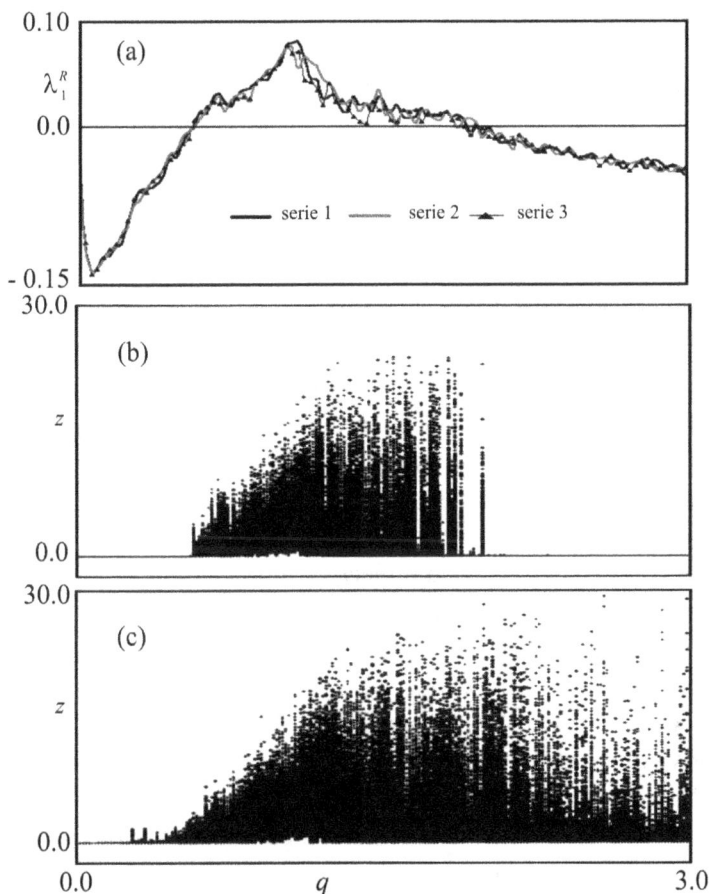

Fig. 4.35. Bifurcation diagram of the largest RLE of system (24.62) with a *non-deterministic drive* (Eqs. (4.64a, b)) estimated using the synchronization method for various periods of estimation T_e: series 1 — $T_e = 5000$, series 2 — $T_e = 10000$, series 3 — $T_e = 30000$ (a); the corresponding *synchronization error* computed according to Eq. (4.43) for $N=5$ identical (b) and slightly different (1% of a mismatch in the parameter α) response oscillators (c). Other parameters of the estimation procedure are $d_{1(0)} = -0.2$, $d_{2(0)} = 0.2$, $\delta = 0.001$, $\varepsilon = 0.01$, $D = 2.0$, $T_t = 0.5 T_e$.

where γ_k and $\varphi(t)$ are random constant and random functions of time, respectively. Hence, the detailed description of the simulated *non-deterministic* signal (for $L=3$) is as follows:

$$\dot{\varphi} = rand[-5,5], \tag{4.64a}$$

$$\begin{aligned} e(t) &= \sin[\Omega t + \varphi(t)] \\ &+ 2\sin[2\Omega t - 2.03\varphi(t)] + 3\sin[3\Omega t + 1.45\varphi(t)], \end{aligned} \tag{4.64b}$$

where Ω represents the dominant frequency of the drive, $\varphi(t)$ is a randomly fluctuating phase, which is under control of the stochastic function *rand* $[-5, 5]$ returning a random number uniformly distributed over $(-5, 5)$. The numerical computations have been carried out employing the RK4 method with a fixed time step $dt = 0.005$. Exemplary time diagrams of the random phase (Eq. (4.64a)) and the corresponding stochastic driving signal (Eq. (4.64b)) are presented in Figs. 4.34a and 4.34b, respectively.

The results of the numerical analysis of the *non-deterministic drive* case are presented in Figs. 4.35a and 4.35b. In Fig. 4.35a bifurcation courses of the largest RLE are depicted. The classical algorithm for calculating Lyapunov exponents is not applicable here due to the stochastic component. Therefore, the largest RLE has been estimated by means of the method proposed in Secs. 5–7. Each of three RLE-curves in Fig. 4.35a has been generated for various conditions the method used (different periods of estimation T_e). Nevertheless, they can be found close to each other on the diagram and they yield almost the same range of the positive RLE. It demonstrates that this method is almost insensitive to the changes in conditions of the estimation process, even in spite of an unrepeatable character of the random drive. The correctness of the determined RLE is verified on the parallel bifurcation diagram of the *synchronization error* (Fig. 4.35b), which has been computed for $N=5$ response oscillators (4.62), according to Eq. (4.43). It is clearly visible that the desynchronous z-interval is corresponding to the positive RLE and *vice versa*. However, like in the previous example of the *deterministic drive*, a slight mismatch of the parameter α (up to 1% over all five response oscillators) destabilizes the synchronous regime (see Fig. 4.35c). A stable state of the ICS can be observed only for lower values of the control parameter q. It is also an analogy to the Lorenz driving case (Fig. 4.33c).

This fact can indicate that the GS observed in the considered arrays of mechanical oscillators is of a weak type in both cases of the deterministic and non-deterministic drive.

4.3.2.2 *Driven Henon maps*

Consider a set of N identical Henon maps with an additional discrete-time signal playing a role of the independent common drive (Eq. (4.61a)). A single i-th response system (Eq. (4.61b)) is described as follows:

$$x_1^* = 1 - ax_1^2 + x_2 + qe,$$
$$x_2^* = bx_1,$$

(4.65)

where a, b are constants. In Eqs. (4.65) and (4.66) the following simplifications in the system description have been assumed: $x_j^* = x_j(n+1)$, $x_j = x_j(n)$ and e is the n-th iteration of the driving signal ($j = 1, 2$).

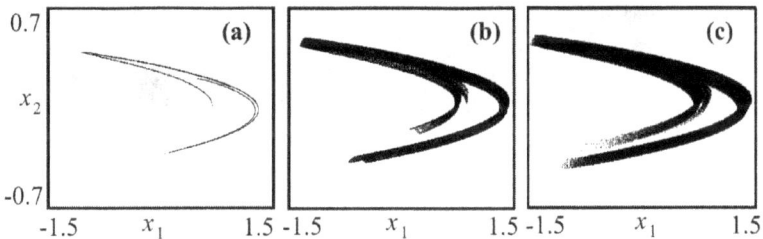

Fig. 4.36. Henon maps: (a) separated (Eq. (4.65)), (b) with the chaotic drive (Eq. (4.66)), (c) with the stochastic drive (Eq. (4.67)) .

In the numerical experiment, $a = 1.2$ and $b = 0.3$ have been taken. For these values of constants, the Henon map (Fig. 4.36a) exhibits chaotic dynamics ($\lambda_1 = 0.307$). The classical logistic map

$$e^* = 3.9e(1-e),$$

(4.66)

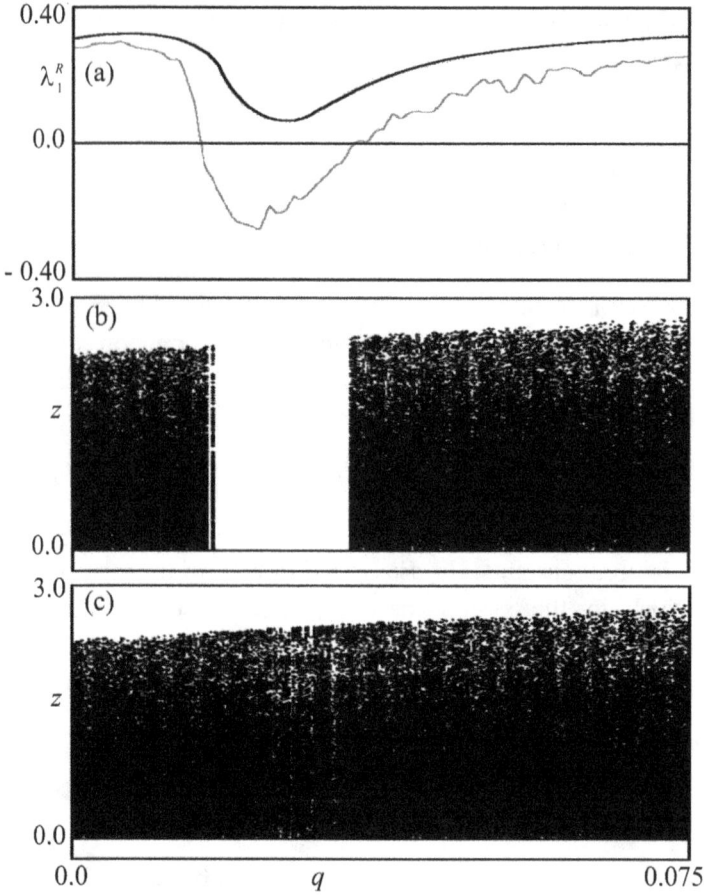

Fig. 4.37. Bifurcation diagram of the largest RLE of the Henon map (Eq. (4.65)) driven by the chaotic logistic map (*deterministic drive*) (Eq. (4.66)) calculated by means of the classical algorithm (in black) and estimated using the synchronization method (in gray) (a); the corresponding synchronization error computed according to Eq. (4.43) for $N = 2$ identical (b) and not noticeably different (10^{-12} of a mismatch in the parameter a) response oscillators (c). Parameters of the synchronization method: $d_{1(0)} = -0.5$, $d_{2(0)} = 0.5$, $\delta = 0.001$, $\varepsilon = 0.001$, $D = 3.0$, $n_e = 100000$, $n_t = 0.5T_e$.

has been applied as the chaotic *deterministic drive* (the positive Lyapunov exponent of system (4.66): $\lambda = 0.495$). On the other hand, the *non-deterministic drive* has been modeled using the stochastic function returning a random number over the range (0, 1), i.e.,

$$e = rand[0,1].$$ (4.67)

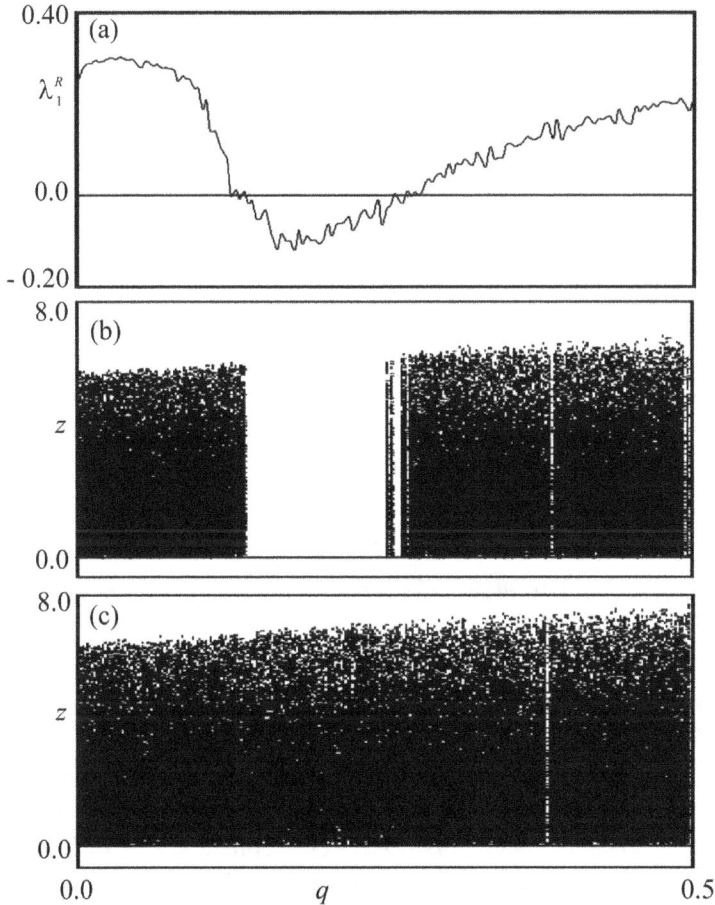

Fig. 4.38. Bifurcation diagram of the largest RLE of the Henon map (Eq. (4.65)) with the *non-deterministic drive* (Eq. (4.67)) estimated using the synchronization method (a); the corresponding *synchronization error* computed according to Eq. (4.43) for $N = 5$ identical (b) and insensibly different (10^{-12} of a mismatch in the parameter a) response oscillators (c). Parameters of the synchronization method: $d_{1(0)} = -0.5$, $d_{2(0)} = 0.5$, $\delta = 0.001$, $\varepsilon = 0.001$, $D = 3.0$, $n_e = 100000$, $n_t = 0.5 n_e$.

Exemplary Henon maps (Eq. (4.65)) with the chaotic (Eq. (4.66)) and stochastic (Eq. (4.67)) drive are shown in Figs. 4.36a and 4.36c, respectively. Their bifurcation analysis performed numerically is depicted in Figs. 4.37a–c and 4.38a–c. The results have been collected and presented in a way analogous to the analysis of driven mechanical oscillators.

In Figs. 4.37a and 4.38a, the largest RLE determined with the synchronization method is presented. Additionally, in Fig. 4.37 the estimated RLE is compared to its equivalent calculated by means of the classical algorithm applied to Eqs. (4.65) and (4.66). A significant discrepancy between both courses of the RLEs occurs, especially in the synchronous interval of q, where the RLE estimated with the proposed method (gray curve) is negative, which indicates the CS of response maps.

However, the RLE computed with the classical algorithm (additionally verified with Ref. Parker & Chua (1989)) is positive in the same interval (black curve), i.e., the CS should not appear. The existence of the synchronous window is confirmed on the corresponding bifurcation diagram of the *synchronization error* (Fig. 4.37b, $N=2$). Thus, we can conclude that the synchronization-based approach has some practical advantages over traditional algorithmic methods with respect to quantifying the synchronizability of dynamical systems.

Similar results have been obtained in case of the stochastic drive (Eq. 4.67). Here the estimated RLE (Fig. 4.38a) also points out the occurrence of the synchronous q-window and this fact is corroborated in Fig. 4.38b for $N=5$ oscillators.

On the other hand, a qualitative difference of the calculated and estimated RLEs (Fig. 4.37a) can indicate that the CS is unstable and sensitive to the parameter mismatch. This conjecture is verified on the bifurcation diagrams shown in Figs. 4.37c and 4.38c. Even for extremely small non-identity of response oscillators (10^{-12} of a mismatch in the parameter a), synchronous windows disappear.

Thus, similarly as in the case of driven mechanical oscillators, the synchronous windows observed in Figs. (4.37b) and (4.38b) are an effect of the weak GS regime.

PART II

APPLICATION

Chapter 5

Lyapunov Exponents: Idea and Calculation

One of the most useful criteria for determining the stability of the synchronous state and dynamical systems are Lyapunov exponents or the Lyapunov function (Lyapunov (1947)). Definition of the Lyapunov exponent was introduced by Russian mathematician V. I. Oseledec (Oseledec (1968)), in form suited for theory of dynamical systems. It describes numbers connected with averaged behaviour of the derivative of mapping along the phase plane trajectory, which is logarithmic measure of sensitivity of a dynamical system to arbitrarily small change of initial conditions. These numbers were named after another Russian scientist A. M. Lyapunov; due to fact their values are qualitative and quantitative illustration of his criterion of stability of dynamical systems (see Sec. 5.1 below).

Considering stability criterions of dynamical systems by Lyapunov (Definitions 5.1 and 5.2), it easy to spot the analogy of them to the definitions of the CS (Eq. (1.3)) and the ICS (Eq. (1.4)), respectively. This analogy suggests close connection between Lyapunov exponents of dynamical systems and processes leading to their synchronization (the CS). This fact was confirmed by the theoretical and numerical analysis presented in Chapters 3 and 4, in Part I of this book. Such relation from one side allows to approximate strength of the coupling at which synchronization occurs (as it was shown in Part I), and, on the other can be used for estimation of the largest Lyapunov exponent in the numerical or sometimes even real experiment.

In this Chapter the general concept of Lyapunov exponents and a review of known methods of their calculation and estimation are presented. Such a compact survey allows us to compare advantages and disadvantages of these classical algorithms with the proposed method based on the above-mentioned properties of the CS, which is demonstrated in detail in Chapters 6 and 7.

5.1 Stability Criterions of Dynamical Systems

Definition 5.1 *The trajectory* $\mathbf{x}(t)$ *is stable in the Lyapunov sense, if for any, arbitrary small* $\varepsilon > 0$, *there exists such* $\delta > 0$ *that for any initial point of the trajectory taken from the neighborhood* $\mathbf{x}(0)$, *i.e.,* $\|\mathbf{x}(0) - \mathbf{y}(0)\| < \delta$, *for all* $t > 0$, *the inequality* $\|\mathbf{x}(t) - \mathbf{y}(t)\| < \varepsilon$ *is fulfilled.*

The above definition is illustrated in Fig. 5.1a. Talking in a less formal way, the dynamical system is stable according to the Lyapunov criterion, if its two phase plane trajectories, which are initiated from very close initial conditions, remain close in the further time evolution of the system. A stronger version of this definition is used in the description of the asymptotic stability (Kuznietsov (2001)).

Definition 5.2 *The phase plane trajectory* $\mathbf{x}(t)$ *is asymptotically stable, when for any, arbitrary small* $\varepsilon > 0$, *there exists such* $\delta > 0$ *that under the condition* $\|\mathbf{x}(0) - \mathbf{y}(0)\| < \delta$, *for all* $t > 0$, *the following relation takes place* $\lim_{t \to \infty} \|\mathbf{x}(t) - \mathbf{y}(t)\| = 0$.

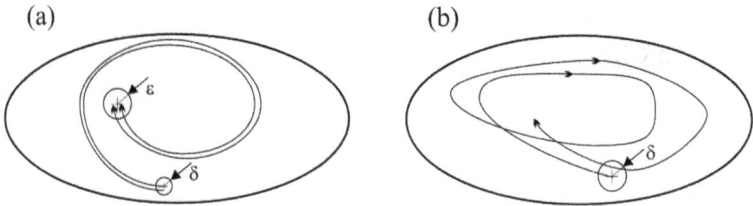

Fig. 5.1. Illustration of an idea of the dynamical system stability in the sense of Lyapunov (a), and Poisson (b).

Additionally, the idea of the Lyapunov exponent is connected also with the definition of stability of a dynamical system in the sense of Poisson (Birkhoff (1927), see Fig. 5.1b).

Definition 5.3 *The phase plane trajectory* $\mathbf{x}(t)$ *is stable in the sense of Poisson, when, and only in the case if it is limited, and, for any values of* $t_0 = 0$ *and* $\delta > 0$, *there exists large enough* t_i ($i = 1, 2, 3, ...; t_0 < t_1 < t_2 < t_3 < ....$), *for which* $\left\| \mathbf{x}(t_i) - \mathbf{x}(0) \right\| < \delta$.

5.2 General Concept of Lyapunov Exponents

A simple description of the Lyapunov exponent is shown in Eq. (5.1). Let the initial distance of the two infinitesimally close trajectories be $\varepsilon(0)$. After a time t, this distance is defined as:

$$\varepsilon(t) = \varepsilon(0)\exp(\lambda t), \qquad (5.1)$$

where λ is the Lyapunov exponent. From Eq. (5.1) it results that close neighboring phase trajectories diverge only when $\lambda > 0$. In almost all cases of dynamical systems, the relation $\lambda \leq 0$ rules out the exponential divergence of near trajectories. Lyapunov exponents can be estimated either for mappings with discrete timing, described with difference equations, or for flows described with differential equations, where time has a continuous character. The exponential convergence and the divergence of neighboring trajectories, according to Eq. (5.1), are exemplified in Figs. 5.2a and 5.2b correspondingly, on the basis of the classical logistic mapping.

Let us consider a dynamical system described by the following equation:

$$\frac{d\mathbf{x}}{dt} = \mathbf{f}(\mathbf{x}, a), \qquad (5.2)$$

where $\mathbf{x}=[x_1, x_2, ... , x_k]^T \in \mathcal{D}$ ($\mathcal{D}$ is an open set in the phase space $\mathfrak{R}^k$), and, $\mathbf{f}=[f_1, f_2, ... , f_k]^T$ is a differentiable function which depends on the parameter a.

The solution to Eq. (5.2) with the initial condition $\mathbf{x}(0)=\mathbf{x}_0$ can be written in the form:

$$\mathbf{x}(t) = \mathbf{T}(t)\mathbf{x}_0 , \tag{5.3}$$

where $\mathbf{T}(t)$ is a mapping describing time evolution of all points in the phase space.

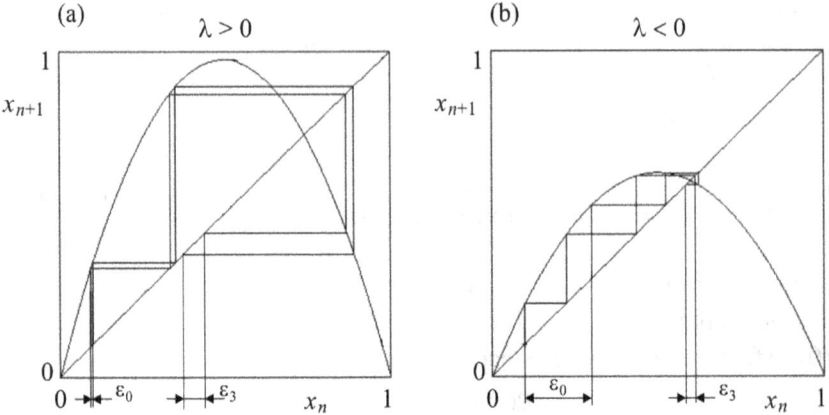

Fig. 5.2. Exponential divergence (a), and, convergence (b) of the two trajectories of the logistic mapping $x_{n+1} = rx_n(1-x_n)$ after three iterations; (a) — r=3.9, (b) — r=1.5.

In the further considerations let us assume that $\mathbf{y}(t)$ is a particular solution to Eq. (5.2). Expanding Eq. (5.2) into a Taylor series in the neighborhood of the particular solution $\mathbf{y}(t)$, and neglecting all terms of the order higher than one, we get:

$$\frac{d\mathbf{x}}{dt} - \frac{d\mathbf{y}}{dt} = \frac{\partial \mathbf{f}[\mathbf{y}(t)]}{\partial \mathbf{x}}[\mathbf{x}(t) - \mathbf{y}(t)]. \tag{5.4}$$

For the analysis of stability of the solution $\mathbf{y}(t)$, let us introduce a variable $\mathbf{z}(t)$ which represents a difference between the disturbed solution $\mathbf{x}(t)$, and the particular one $\mathbf{y}(t)$:

$$\mathbf{z}(t) = \mathbf{x}(t) - \mathbf{y}(t). \tag{5.5}$$

Substituting Eq. (5.5) into Eq. (5.4), we obtain a linearized equation in the form:

$$\frac{d\mathbf{z}}{dt} = \mathbf{J}(\mathbf{y}(t))\mathbf{z} , \tag{5.6}$$

where

$$J(y(t)) = \frac{\partial f[y(t)]}{\partial x}$$

is the Jacobi matrix defined in the point $y(t)$. Such a linearization in the neighborhood of the particular solution was performed in the way similar to a linearization in the neighborhood of the critical point. One should remember that in the discussed case, both $y(t)$ and $J(y(t))$ are not a constant function or a matrix with constant coefficients, respectively.

Let us assume that $y(0)=y_0$ is an initial point (for $t=0$) of the solution $y(t)$, and, $z(0)=z_0$ is an initial value of the disturbance introduced to the system. There exists a fundamental set of solutions for Eq. (5.6), which is composed of k linearly independent solutions of the equation:

$$\frac{dZ(t,y_0)}{dt} = J(y(t))Z(t,y_0) . \tag{5.7}$$

As a result, the solution of Eq. (5.6) may be described as:

$$z(t) = Z(t,y_0)z_0 , \tag{5.8}$$

where $Z(t, y_0)$ is a fundamental matrix of solutions to Eq. (5.6).

Definition 5.4 *The Lyapunov exponent of linearized Eq. (5.6) is the number defined as:*

$$\lambda_i = \lim_{t \to \infty} \frac{1}{t} \ln \left\| z^i(t, y_0, e^i) \right\|, \tag{5.9}$$

where $z^i(...)$ is the i-th (i=1, 2, 3, ...,k) fundamental solution of the system (5.6), e^i — i-th unit vector, and, $\|...\|$ stands for any norm in the space $\Re^k$ (the Lyapunov exponent is independent of a choice of the norm). This exponent can be also defined as:

$$\lambda_i = \lim_{t \to \infty} \frac{1}{t} \ln |m_i(t)|, \tag{5.10}$$

where $m_i(t)$ are eigenvalues of solution (5.8).

Equation (5.9) defines the Lyapunov exponent in the direction of the unit vector e (e^i). There exists a k-number of such independent vectors, so one can choose k of the base vectors. The above definition fully characterizes all possible Lyapunov exponents due to fact that the fundamental solution linearly depends on e. To show a relation of Eqs. (5.9) and (5.10), one should note the following:

$$\|z(t, y_0, e)\|^2 = e^T Z^T Z e. \tag{5.11}$$

If we have a symmetrical quadratic form calculated on the vector e on the right-hand side of Eq. (5.11), then in the case when e is an eigenvector of the symmetrical matrix $Z^T Z$, we get an absolute squared eigenvalue. This way the Lyapunov exponents can be characterized with their eigenvalues according to formula (5.10).

For discrete-time dynamical systems described with difference equations of the general form:

$$\mathbf{x}_{n+1} = \mathbf{g}(\mathbf{x}_n), \tag{5.12}$$

where $x \in \Re^k$ and $g=[g_1, g_2, \dots, g_k]^T$, the solution obtained after n steps may be accounted using the initial condition x_0 iterating n times the mapping g. Then, the Jacobi matrix $J_n(x_0)$ of the iterated mapping is a product of the Jacobi matrix of the mapping g calculated in the points related to the previous iterations relatively, according to the following formula:

$$\mathbf{J}_n(\mathbf{x}_0) = \mathbf{J}(\mathbf{x}_{n-1})\mathbf{J}(\mathbf{x}_{n-2})\dots\mathbf{J}(\mathbf{x}_0) = \prod_{r=0}^{n-1} \mathbf{J}(\mathbf{x}_r), \tag{5.13}$$

where $J(x_r) = \partial g_i / \partial x_j$; $i, j = 1, 2, 3, \dots k$. Concluding, the Lyapunov exponents of such mappings can be defined with a relation similar to Eq. (1.19).

Definition 5.5 *Lyapunov exponents of the mapping* (5.12) *are numbers defined as*

$$\lambda_i = \lim_{n \to \infty} \frac{1}{n} \ln |\sigma_i(n)|, \tag{5.14}$$

where $\sigma_i(n)$ *are eigenvalues of the matrix* $\mathbf{J}_n(\mathbf{x}_0)$ (*Eq.* (5.13)).

The above Eqs. (5.9), (5.10) and (5.14) define an idea of the so-called one-dimensional Lyapunov exponents. A more general concept of the Lyapunov exponent is described with the following definition (Shimada, Nagashima (1979)).

Definition 5.6 *A number defined with the formula*:

$$\lambda_i(e^m, \mathbf{y}_0) = \lim_{t \to \infty} \frac{1}{t} \ln \frac{\left\| \mathbf{Z}(t, \mathbf{y}_0) e_1 \wedge \mathbf{Z}(t, \mathbf{y}_0) e_2 \wedge \ldots \wedge \mathbf{Z}(t, \mathbf{y}_0) e_m \right\|}{\left\| e_1 \wedge e_2 \wedge \ldots \wedge e_m \right\|} \quad (5.15)$$

is called an m-dimensional Lyapunov exponent (for m=1, 2, 3, ... , k).

The symbols in the above formula have the following meanings: e^m is the k-dimensional subspace tangent in the point y_0 to the phase manifold, $\{e_i\}$ (i=1, 2, 3, ... , m) is a set of base axes of the subspace e^m, $\wedge$ stands for an external product, and $\|...\|$ is any norm in the space $\mathfrak{R}^m$. The number defined by formula (5.15) represents an exponential measure of the change of volume of the element of the subspace e^m during evolution along the orbit starting in the point y_0. It can be noted that for $m = 1$, formula (5.15) is reduced to the form given by Eq. (5.9).

The main properties of such a definition of Lyapunov exponents are as follows:

1. The one-dimensional exponent $\lambda(e^1, \mathbf{y}_0)$ can take no more than k different values.
2. The m-dimensional exponent $\lambda(e^m, \mathbf{y}_0)$ can take no more than C_m^k different values, and every one of them is related to the sum of m one-dimensional Lyapunov exponents, according to the formula:

$$\lambda(e^m, \mathbf{y}_0) = \sum_{i=1}^{m} \lambda(e^i, \mathbf{y}_0). \quad (5.16)$$

   For example, for k=3, m-dimensional exponents can take the following values:
   $\lambda(e^1, \mathbf{y}_0)$ = one of the numbers $\{\lambda_1, \lambda_2, \lambda_3\}$,
   $\lambda(e^2, \mathbf{y}_0)$ = one of the numbers $\{\lambda_1+\lambda_2, \lambda_1+\lambda_3, \lambda_2+\lambda_3\}$, or,
   $\lambda(e^3, \mathbf{y}_0)$ = $\{\lambda_1+\lambda_2+\lambda_3\}$.
3. If the set of the base axes $\{e_i\}$ (i=1, 2, 3, ... , m) of the tangent subspace is chosen in a random way, then the m-dimensional exponent $\lambda(e^m, \mathbf{y}_0)$ (m=1, 2, 3, ... , k) is convergent to the maximal value of the set of C_m^k different numbers, relatively, with probability equal to 1.

It is obvious that the Lyapunov exponents exist and that they are limited only in the case when a limit in Eqs. (5.9), (5.10), (5.14) and (5.15) can be found. Existence of such a limit has been proven by Oseledec (1968) who has introduced this idea of the Lyapunov exponent as a criterion of an assessment of the quality of motion of the dynamic system.

Theorem 5.1 (Oseledec (1968)) *If there exists* $\mathbf{T}(t)$ *(Eq. (5.3)) and an invariant measure* μ *and also* $\|\partial\mathbf{f}/\partial\mathbf{x}\| \in L^1(\mu)$, *then* m-*dimensional Lyapunov exponents* $\lambda(e^m, \mathbf{y}_0)$ *(m=1, 2, 3, ...,k) exist for almost all* $\mathbf{y}_0$ (cited after Ref. Shimada, Nagashima (1979)).

In practice the term "Lyapunov exponents" refers to its one-dimensional case and this rule is also assumed in this book.

As a result, concluding the above-mentioned considerations, one can state that for the dynamical system given by Eq. (5.2) there exists a set of k Lyapunov exponents, i.e., their number is equal to the dimension of the phase space of the system. Such a set $\{\lambda_i\}$ (i=1, 2, 3, ... , k) being ordered according to the less-equal relation is called a *spectrum of the Lyapunov exponents* of the trajectory $\mathbf{y}(t)$. A simplified form of the set $\{\lambda_i\}$ is the *spectrum of the signs of Lyapunov exponents*, defined as an ordered set of symbols: +, 0, –, which are related to the positive, zero or negative values of the exponents λ_i. From properties of the Lyapunov exponents, one can conclude that the sum of all of them (the k-dimensional Lyapunov exponent, according to formula (5.16)) is equal to the divergence of the phase flow (Anishchenko (1990)). Then, the volume of the phase space $V(t)$, in which the disturbed solution $\mathbf{x}(t)$ evolves in the neighborhood $\mathbf{y}(t)$, can be formulated as:

$$V(t) = V(0)\exp\left(t\sum_{i=1}^{k}\lambda_i\right),\qquad(5.17)$$

where $V(0)$ is an initial volume of the phase space. From Eq. (5.17) and the theorem of divergence (Schuster, Just (2006)), one can define *dissipative dynamical systems* as such, for which:

$$\sum_{i=1}^{n} \lambda_i < 0, \tag{5.18}$$

whereas the above sum is equal to zero for any conservative system.

Usually, in literature both the Lyapunov exponents and the eigenvalues are marked with the symbol λ, due to the fact that the exponents are a generalization of the eigenvalues of the linearized matrix in the stable point. According to that, when all of the Lyapunov exponents are negative ($\lambda_i < 0$), the solution of Eq. (5.2) is a critical point, being actually an attractor. The Lyapunov exponents are related to the real parts of the eigenvalues of system (5.2) in this point. In the case of the periodic solution, when a stable limit cycle is an attractor, the maximal exponent is equal to zero ($\lambda_1 = 0$, $\lambda_i < 0$, $i=2, 3, \ldots, k$), while an attractor of the quasi-periodic solution is a torus, which is characterized with a pair of the zero valued maximum Lyapunov exponents ($\lambda_1 = 0$, $\lambda_2 = 0$, $\lambda_i < 0$, $i=3, 4, \ldots, k$). In such cases the solutions $\mathbf{y}(t)$ are stable, both in the Lyapunov and Poisson sense. When the largest Lyapunov exponent is positive ($\lambda_i > 0$) and simultaneously inequality (5.18) is fulfilled, then the trajectory $y(t)$ for all $t > 0$ evolves in a set bounded in the phase space (a compact set), usually called a strange, chaotic attractor. Such a set of trajectories is defined by instability in the Lyapunov sense, but it is stable in the Poisson sense, due to the bounded solution of Eq. (5.2).

Another important property of the spectrum of Lyapunov exponents (from a viewpoint of applications) is its usefulness for an estimation of the fractal dimension of strange attractors. This idea of the fractal dimension d^F, introduced by Mandelbrot (1982), generalizes the concept of k-dimensional volume and has a strictly metric sense. In agreement with the Kaplan–Yorke conjecture (Kaplan, Yorke (1979)), the fractal dimension d^F can be approximated by the Lyapunov dimension d^L calculated from the spectrum of Lyapunov exponents, written in order $\lambda_1 > \lambda_2 > \ldots > \lambda_k$, according to the formula:

$$D_L = j + \frac{\sum_{i=1}^{j} \lambda_i}{\left|\lambda_{j+1}\right|}, \tag{5.19}$$

where j is the largest integer number for which the following inequality is fulfilled:

$$\sum_{i=1}^{j} \lambda_i \geq 0 .$$

In general, main types of the attractors of dissipative dynamical systems (fulfilling relation (5.18)) can be classified using a spectrum of the signs of Lyapunov exponents according to the following scheme:

Table 5.1. Spectrum of Lyapunov exponents and corresponding types of attractor.

Spectrum of Lyapunov exponents	Attractor type
$(-, -, -, \ldots, -)$	critical point,
$(0, -, -, \ldots, -)$	limit cycle,
$(0, 0, -, \ldots, -)$	torus,
$(+, 0, -, \ldots, -)$	strange chaotic attractor,
$(+, \ldots, +, 0, -, \ldots, -)$	strange hyperchaotic attractor.

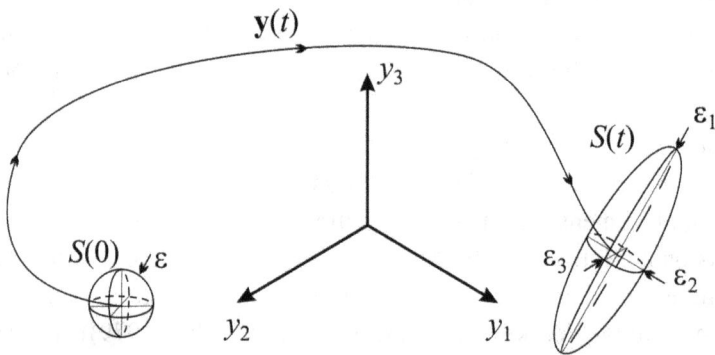

Fig. 5.3. Geometrical interpretation of Lyapunov exponents.

The geometrical interpretation of Lyapunov exponents for $k=3$ is presented in Fig. 5.3. During the time evolution of the system, an initial sphere $S(0)$ with an infinitesimally small radius ε, surrounding the point

$\mathbf{y}(t_0)$ of the phase plane trajectory lying on the strange attractor, deforms to an ellipsoid $S(t)$ as a result of a change of its main axes in a proportional way to the values of the Lyapunov exponent connected to them. This way, after the time $t–t_0$, the length of the i-th main axis of the ellipsoid reaches ε_i, according to formula (5.1), and, the i-th Lyapunov exponent is described as:

$$\lambda_i = \frac{1}{t-t_0} \ln \left| \frac{\varepsilon_i}{\varepsilon} \right|. \tag{5.20}$$

The Lyapunov exponents are characterized by 'divergent properties' of the vector field $\mathbf{f}$ in eigendirections of the Jacobi matrix $\partial \mathbf{f}/\partial \mathbf{x}$. Because the orientation of the main axes of the above-described ellipsoid (see Fig. 5.3) changes during the time evolution of the system, the directions related to each exponent also change during the "journey along the attractor". That is why we cannot speak about any well-defined direction of the phase space, which could be connected to the particular Lyapunov exponent. But one of the main axes always shows a direction tangent to the phase space trajectory. In this particular direction, the convergence or the divergence of the points lying on the same orbit is not possible at all, so all of the attractors, which contain one or more trajectories, are characterized by the spectrum of the exponents having a zero value, which describes the "divergence" of the phase flow in the direction tangent to the orbit. In the case of mappings, such a trajectory is a discrete set of points, so no tangent to it exists. Then, in the spectrum of the Lyapunov exponents of the discrete map, the zero value can appear only in the case of a quasi-periodic solution.

5.3 Calculation of Lyapunov Exponents

The analytical determination of Lyapunov exponents is possible only in the considerations of simple dynamical systems. Owing to this fact, in most cases numerical methods are used to calculate their values. There is a difference in a way of calculation of the exponents for discrete-time maps (λ_m) and phase flows (λ_f), which is a result of different forms of the transformation from one step to another for flows (integration) and maps (simple mapping of the previous iteration). To explain the difference, let us discuss some examples of a simple dynamical system, described with difference equations first, and then with the differential ones (Schuster, Just (2006)). The Lyapunov exponent for the mapping:

$$x_{n+1} = ax_n$$

equals $\lambda_m = \ln a$, because $x_n - y_n = (x_0 - y_0)\exp(n \ln a)$, while for the flow it is

$$\dot{x} = ax,$$

and we can conclude that the neighboring trajectories diverge at a speed a, i.e., the exponent $\lambda_f = a$, because $x(t) - y(t) = (x_0 - y_0)\exp(at)$.

   Later in this work, some compact discussions of frequently used numerical algorithms for calculation and estimation of Lyapunov exponents are presented.

5.3.1 *Classical algorithms*

The first numerical characteristics of chaotic behavior of the dynamical system, which presented a divergence of neighboring trajectories, was conducted by Henon and Heiles (1964). An efficient algorithm for calculating the complete *spectrum of the Lyapunov exponents* based on the Oseledec theorem was independently formulated both by Benettin *et al.* (1976) and Shimada and Nagashima (1979). Then, it was improved by Benettin *et al.* (1980a, 1980b) and Wolf (1986). The Lyapunov exponents of discrete-time system (5.12) can be determined in a numerical way by calculating eigenvalues of Jacobi matrix (5.13) for successive steps of the mapping, and substituting them into Eq. (5.14). In the case of phase flows, to determine their values, Eq. (5.10) can be used after the integration of equation of motion (5.2) and linearized Eq. (5.6), with typical (generic) values of the initial conditions. But in practice, a direct application of theorem 5.1 to the numerical estimation of the Lyapunov exponent is not possible. After some time of calculations, lengths of the eigenvectors of the Jacobi matrix become too large, and angles between them are consecutively too small to continue the calculations. This happens because for the dissipative systems an initial volume of the element of the subspace tangent to the phase manifold e^m

(parallelepiped), defined with the vector product of the initially orthogonal eigenvectors of the matrix $\mathbf{J(y)}$, converges to zero due to a dissipative convergence (in the consecutive iterations) of the directions of such vectors related to the vector with the highest increase rate, as shown in Fig. 5.4.

To avoid troubles, one can use a procedure of re-orthonormalization by Gram–Schmidt (GSR). After determination of the base of the orthonormal vectors $\{u_i\}$ (i = 1, 2, 3, ..., k, where u_i is an approximation of the direction of the i-th axis of the sphere $S(t)$ for $\mathbf{y}=\mathbf{y}(t_n)$) and the definition of the vector $v_i=\mathbf{J(y)}u_i$ the Gram–Schmidt algorithm for calculation of the orthogonal vectors w_i can be used in the following manner. The vector w_1 is equal to the vector v_1. Let us assume that the vectors w_1, w_2, ..., w_{i-1} for $i > 1$ are determined. Labeling the scalar product with $<\cdot,\cdot>$ and the Euclid's norm with, we obtain:

$$w_i = v_i - \sum_{j=1}^{i-1} <v_j,u_j^*> u_j^*, \qquad (5.21)$$

where u_j^* is a unit vector $w_j / \|w_j\|$. If a vector w_j is a zero one, then the calculations stop. This way the vectors u_j^* create a new set of orthonormal vectors, and the length $r_i = \|w_j\|$ determines an increase of the ellipsoid in the i-th direction.

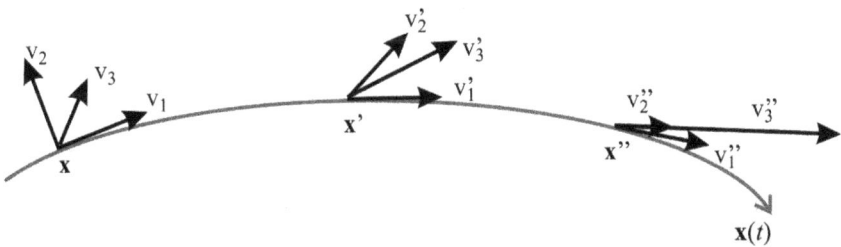

Fig. 5.4. Three-dimensional visualization (k=3) of the convergence of the eigenvectors of the matrix $\mathbf{J(y)}$ related to the direction of the vector with the largest increase (v_3).

The vectors u^*_j determine approximate directions of the i-th axis of the ellipsoid $S(t_{n+1})$ in the next iteration, and r_i is an approximate ratio of the i-th axis of the ellipsoid $S(t_{n+1})$ to the i-th axis of the ellipsoid $S(t_n)$ (taken from the previous iteration). Such calculations are conducted in every consecutive iteration of the investigated system. The value $r_{i,j}$ labels a value of the j-th iteration. Then,

$$\lambda_i(t_n) = \frac{1}{n} \sum_{i=1}^{n} \ln r_{i,j} \qquad (5.22)$$

is an approximate value of the i-th Lyapunov exponent, which converges to λ_i with an increase of n.

The above-presented numerical method (Nusse, Yorke (1994)) allows for the calculation of the complete spectrum of the k 1D Lyapunov exponents both for maps and for phase flows. Such a spectrum can be calculated by using the re-orthonormalization procedure (GSR) for calculation of the k consecutive largest m-dimensional Lyapunov exponents $\lambda_1(e^m)$ ($\lambda_1(e^1)$, $\lambda_1(e^2)$, ... , $\lambda_1(e^k)$), and then for calculation of further 1D exponents using the following formula (Benettin *et al.* (1976), Shimada and Nagashima (1979)):

$$\lambda_i = \begin{cases} \lambda_1(e^m) & \text{dla} \quad i = m = 1 \\ \lambda_1(e^m) - \lambda_1(e^{m-1}) & \text{dla} \quad i = m > 1 \end{cases}. \qquad (5.23)$$

Such a numerical algorithm can be successfully used for many dynamical systems, both autonomic and non-autonomic. But a range of its use is limited by the condition of continuity of the system equation of motion (5.2) in the complete time range analyzed. The discontinuity in the above equations makes linearization of the system according to Eqs. (5.4) – (5.7) impossible.

5.3.2 *Evaluation of the Lyapunov exponents from the time series*

The above-described algorithms allow us to determine a spectrum of the Lyapunov exponents of the analyzed dynamical systems or only the largest of them, provided the equations of motion are explicitly known. Known also is a method of estimation of the Lyapunov exponents based on the scalar time signal, which can be used for both time series generated from the equations of motion and the experimental series acquired without any knowledge of the equations of motion. The method is based on a procedure of reconstruction of the attractor in the phase space from the time dependence of one of the coordinates that was introduced by Takens (1981). The first numerical algorithm for estimation of the largest Lyapunov exponent, derived from it, was formulated by Wolf *et al.* (1985). The time history of the variable $y(t)$ allows for reconstruction of the m-dimensional phase portrait by using the delayed coordinates in such a way that the point on the attractor is described by a set of coordinates $\{y(t), y(t+\tau), y(t+(m-1)\tau)\}$, where τ is an arbitrary chosen time delay. The choice of the initial point $\{y(t_0), y(t_0+\tau), y(t_0+(m-1)\tau)\}$ and then of an analogous point on the neighboring trajectory allows us to determine an initial distance between the points $l(t_0)$. Because the attractor is reconstructed from a single trajectory, as a close point lying on the same attractor, it can be treated as a point of this trajectory, but shifted in time by at least one orbit period. After a time t_1, the distance between these points is equal to $l'(t_1)$. Then, the initial distance $l(t_{i-1})$ and the distance $l'(t_i)$ is determined again in the following iterations after the time t_i ($i=1, 2, 3, ...,n$), as shown in Fig. 5.5a. As a result, the largest Lyapunov exponent can be estimated according to the formula:

$$\lambda_1 = \frac{1}{t_n - t_0} \sum_{i=1}^{n} \ln \frac{l'(t_i)}{l(t_{i-1})} . \tag{5.24}$$

One can easily see that the procedure is similar to the above-presented scheme of calculating λ_1, but, in fact, it is more complicated in practical realization. The method allows one to estimate further non-negative, one-dimensional Lyapunov exponents, if they exist for the system, by estimating values of the consecutive sums of m exponents ($m=1, 2, j$, where j is the largest number of non-negative exponents in the spectrum), i.e., the m-dimensional Lyapunov exponent, according to formula (5.24), expanded into m dimensions, given by:

$$\sum_{m=1}^{j} \lambda_i = \frac{1}{t_n - t_0} \sum_{i=1}^{n} \ln \frac{A'_{m+1}(t_i)}{A_{m+1}(t_{i-1})} , \tag{5.25}$$

where $A_{m+1}(t_i)$ i $A'_{m+1}(t_{i-1})$ are elements of the volume of the phase space determined by $m+1$ points from neighboring orbits. The estimation of the sum of the exponents according to formula (5.25) is illustrated in Fig. 5.5b for $m=2$ (an element of the volume of the phase space is in this case a triangle). Then, one can determine values of the one-dimensional exponents according to Eq. (5.23). But the use of the method is limited by some criteria, which have to be fulfilled in the numerical realization (a wider discussion is to be found in Ref. Wolf *et al.* (1985)). Later publications include a numerical modification of the algorithm (Kantz (1994), Rosenstein *et al.* (1993)).

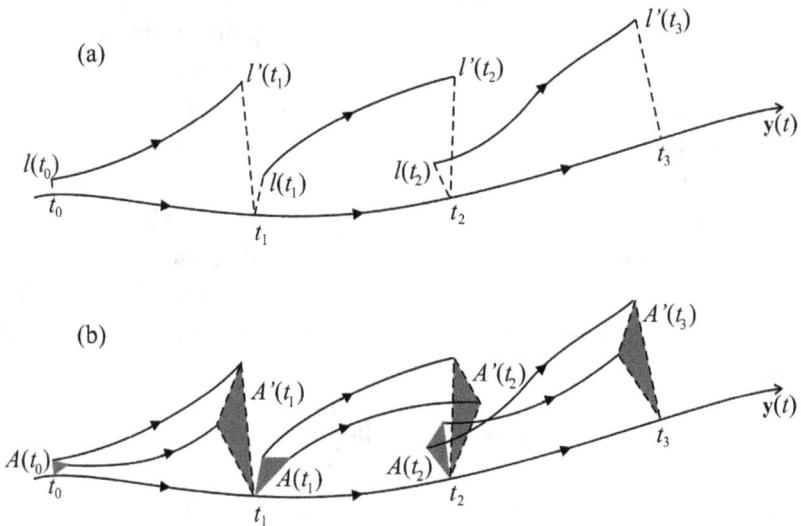

Fig. 5.5. Estimation of the maximum exponent (a), and a sum of two largest Lyapunov exponents (b), based on the scalar time history.

The expanded version of the method which allows for estimation of a complete spectrum of the Lyapunov exponents (including the negative ones) can be found in the publication by Sano and Sawada (1985), Eckmann *et al.* (1986), Stoop and Meier (1988). However, in such a spectrum there appear also the so-called spurious Lyapunov exponents (Parlitz (1992)). Hence, the problem here is to distinguish between the true and spurious exponents. In practice, this attempt uses approximate methods for rough estimation of the fundamental matrix of solutions to Eq. (5.6) for the analyzed time history, i.e., the determination of a solution to the equation. Such an approximation can be performed using for example an error estimation procedure by smallest squares. Then, the Lyapunov exponents can be calculated using a formula analogous to Eq. (5.9).

The method can be applied in the case of systems with discontinuities. But in general, there exists an opinion among the researchers that such estimation is too heuristic in its nature. This follows from the fact that one cannot be absolutely sure about the nature of the attractor generated from the time series holding only one coordinate to reflect fully properties of the original attractor (Schuster, Just (2006)).

5.3.3 *Methods for non-differentiable systems*

In the real world there are many dynamical systems with discontinuities. Among them mechanical systems with impacts, dry friction or piecewise linear stiffness characteristics are of most importance. One can add electrical oscillators with partially linear resistance characteristics, for example, a Chua oscillator (Chua *et al.* (1986), Chua (1993), (1994)). In some cases, the nonlinearity in equations can be "smoothed" for the purpose of adaptation of the above described (Sec. 5.3.1) algorithms for calculation of the Lyapunov exponents for the continuous systems. This is usually done by an approximation of the formula describing the nonlinearity with an appropriate continuous function. An example of such an attempt can be a replacement of the function defining the dry friction force T in the Coulomb model with a continuous function "$\tan^{-1}$", according to the following formula (Oestreich (1998)):

$$T = \begin{cases} Nf & \text{dla} \quad v_r \geq 0 \\ -Nf & \text{dla} \quad v_r < 0 \end{cases} \approx \lim_{a \to \infty} Nf \frac{2}{\pi} \arctan(av_r),$$

where N — normal force, f — a dry friction coefficient, v_r — relative velocity, a — a parameter. But there is always danger that the "smoothed" system can be too simplified as compared to the original one.

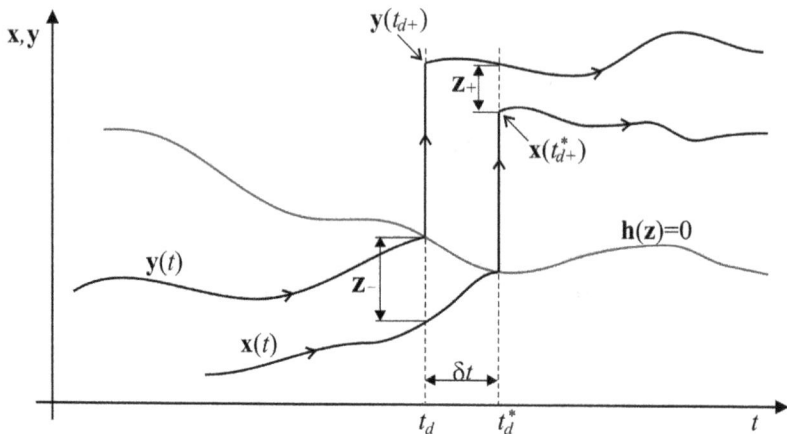

Fig. 5.6. Graphical illustration of the passing of the non-disturbed trajectory $\mathbf{y}(t)$ and the disturbed one $\mathbf{x}(t)$ through the discontinuity.

Over a dozen years ago a method of the calculation of the Lyapunov exponents based on Oseledec theorem 5.1 expanded into dynamical systems with discontinuities was proposed (Müller (1995)). In such cases, the linearization of the equations of motion must be accompanied by a clear statement of the conditions and the transition functions while the trajectory is passing through the discontinuity. The equations with discontinuity can be in the vicinity (time t_d) described by the following form:

$$t_0 \leq t < t_d: \qquad \dot{\mathbf{y}} = \mathbf{f}_1(\mathbf{y}), \qquad \mathbf{y}(t_0) = \mathbf{y}(0), \qquad (5.26a)$$

$$t = t_d: \qquad 0 = \mathbf{h}(\mathbf{y}(t_d)), \qquad (5.26b)$$

$$\mathbf{y}(t_d +) = \mathbf{g}(\mathbf{y}(t_d -)), \qquad (5.26c)$$

$$t > t_d: \qquad \dot{\mathbf{y}} = \mathbf{f}_2(\mathbf{y}), \qquad \mathbf{y}(t_d) = \mathbf{y}(t_d +), \qquad (5.26d)$$

where $f_1(y)$, $f_2(y)$ are equations of motion just before (sign "−") and just after (sign "+"), respectively, the moment of contact of the phase trajectory with the discontinuity. The function h(y) defines the moment of contact, and it may be a scalar function, for example in the case of impacts or a vector sliding with dry friction. Eq. (5.26c) defines the conditions of passing through the discontinuity, for instance, by a restitution coefficient in the case of impacts. It is assumed that all functions in Eqs. (5.26a–d) are at least once differentiable in the time domain.

Let us consider the behavior of the neighboring trajectories, one of which ($\mathbf{x}(t)$) being disturbed (Fig. 5.6). A difference between them is described with the variable $\mathbf{z}(t)$ (Eq. (5.5)). The disturbed trajectory contacts the discontinuity in the time $t_d^* = t_d + \delta t$, so it evolves according to the equations analogous to Eqs. (5.26a–d), but in the time equal to t_d^*. Accordingly, linearized Eq. (5.6) takes the form:

$$t_0 \leq t < t_d: \qquad \dot{\mathbf{z}} = \mathbf{F}_1(t)\mathbf{z}, \qquad \mathbf{z}(t_0) = \mathbf{z}(0), \qquad (5.27)$$

where $F_1(t)$ is the Jacobi matrix of the function $f_1(y)$. The condition of passing for Eq. (5.27) into a linearized form of Eq. (5.26d) is as follows:

$$t = t_d:$$

$$\mathbf{z}_+ = \mathbf{G}(\mathbf{y}_-)\mathbf{z}_- - [\mathbf{G}(\mathbf{y}_-)\mathbf{f}_1(\mathbf{y}_-) - \mathbf{f}_2(\mathbf{y}_+)]\frac{\mathbf{H}(\mathbf{y}_-)}{\mathbf{H}(\mathbf{y}_-)\mathbf{f}_1(\mathbf{y}_-)}\mathbf{z}_-, \qquad (5.28)$$

where $H(y_-)$ and $G(y_-)$ are Jacobi matrices for the functions h(y) and g(y), respectively (the detailed way how to obtain formula (5.28) can be found in Ref. Müller (1995)). Linearizing Eq. (5.26d) after the substitution of condition (5.28), we arrive at the linearized equation of motion after the trajectory passes through the discontinuity:

$$t > t_d: \qquad \dot{\mathbf{z}} = \mathbf{F}_2(t)\mathbf{z}, \qquad \mathbf{z}(t_d) = \mathbf{z}_+, \qquad (5.29)$$

where $F_2(t)$ is the Jacobi matrix of the function $f_2(y)$.

Equations (5.27–5.29) present a generalized procedure of the "linearization" of the dynamical systems with discontinuities, after which the classical algorithm for calculation of the spectrum of Lyapunov exponents can be performed. But, as shown in Fig. 5.6, the algorithm becomes complicated in comparison to the continuous systems, due to a necessity to initialize the equations of motion from the very beginning after the trajectory passes through the discontinuity.

Practically, the value of the maximum Lyapunov exponent is a satisfying criterion of a quality of the system motion and its measure of the sensitivity to initial conditions is maximum. In the subject literature one can find a few examples of algorithms for calculation of its value, and some of them may be used also in systems with discontinuities. A scheme of the simplest one (Benettin *et al.* (1980a) and (1980b)) is presented in Fig. 5.7. To obtain the value λ_1 it is not necessary to integrate Eq. (5.6), so there is no need to use the GSR procedure. It is enough to trace an evolution of two close orbits, which are initially separated by a distance $\mathbf{z}(0)$ in the time τ, while both $\mathbf{z}(0)$ and τ, respectively, should be small enough, for linear effects to remain dominating. To avoid any overflow of the maximum values of the numbers which are available in the particular machine, one should rescale the ratio $\mathbf{z}(i\tau)/\mathbf{z}(0)$ to the value equal to 1 after every i-th iteration ($i=1, 2, 3, ..., n$, where n is the overall number of iterations, see Fig. 5.7). The value λ_1 after the time $t_c=n\tau$ can be expressed as:

$$\lambda_1 = \frac{1}{n\tau}\sum_{i=1}^{n}\ln\left|\mathbf{z}(i\tau)\right|. \tag{5.30}$$

There exists also a method for determining the dominant Lyapunov exponents, which is based on the reduction of the phase flow into a binary series. This auto-regressive method of estimation of the so-called macroscopic Lyapunov exponent λ_{mc} (Singh & Joseph (1989)) is based on binary series, which are usually obtained from the properties of the system trajectories. Such an attempt is mainly applied in the case of specific "two-wing" attractors, as those of Lorenz or Duffing types, where the trajectory oscillates about two unstable fixed points, symmetrically situated in the phase space. The binary sequence contains values of $+1$ or -1, depending on the point around which the trajectory makes an actual loop. Such a binary series holds some information about its original system, and the macroscopic Lyapunov exponent is determined on the basis of a distance (in the successive iterations) between the points obtained from the crossings of the phase plane trajectories with the plane containing the critical points.

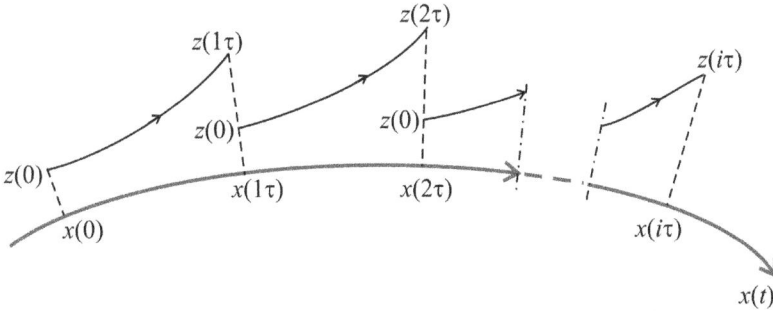

Fig. 5.7. Numerical calculation of the largest Lyapunov exponent.

Another class of methods for the LEs calculation employs reduction of the dynamics of the phase flow determined in the k-dimensional phase space to a l-dimensional discrete map $\mathcal{M}$, for example, a Poincaré map (Poincaré (1913)) or an impact map, $l=1, 2, 3, ...,k-1$, a local map (Jin *et al.* (2006)), etc.. Then, the Lyapunov exponents of such a mapping are determined using Eqs. (5.13) and (5.14). First examples of such an approach can be found in Refs. Hilborn (1994), Oestreich *et al.* (1996), Oestreich (1998). The main application problem here lies in defining the Jacobi matrix of the mapping, where consecutive iterations are not explicitly defined by the known difference equation (Eq. (5.12)) but they are reconstructed from the flow. The newer example of such a map-based approach is the method by Galvanetto (2000), who applied implicitly defined maps for calculation of the two largest LEs of the 2-DoF stick–slip system or the algorithm for impact oscillators formulated by Souza & Caldas (2004), which exploits the so-called transcendental maps that take into account the solution to integrable differential equations, between impacts, supplemented by transition conditions at the instants of the impact. On the other hand, most recently, Jin *et al.* (2006) focused on the local mappings of non-smooth systems and found a general calculation method of the spectrum of Lyapunov exponents for k-dimensional non-smooth dynamical systems, which could be generally applied to the systems with rigid or flexible constraints.

The largest Lyapunov exponent can be estimated also from a one-dimensional mapping of the dynamics of the system — the so-called

H-map, without defining the Jacobian, given by Eq. (5.13) (Oestreich *et al.* (1996), Popp *et al.* (1996)). The idea of such an attempt is calculation of the value of λ_1 on the basis of the averaged value of the directional coefficient of a line tangent do the H-map in the point of its contact with the phase plane trajectory (Hinrichs *et al.* (1996)).

One should remember that the discrete mapping of the phase flow leads to a reduction of the spectrum of the Lyapunov exponents by a number of $k-l$ components, but always at least by a zero exponent, which is connected to the direction tangent to the trajectory. The dependence of the Lyapunov exponent of the phase flow λ_f and the analogous one, which is derived from its mapping λ_m, can be approximated in the form:

$$\lambda_m \approx \lambda_f T_A , \qquad (5.31)$$

where T_A is an averaged time between the successive mappings of the phase flow.

Chapter 6

Determination of the LLE Using the Complete Synchronization

In this Chapter a theoretical background for the LLE estimation numerical procedure is explained. This method can be consider as an application of the properties of the CS via the *diagonal coupling* clarified in detail in Chapter 3 (Sec. 3.2) and demonstrated by some examples in Chapter 4 (Sec. 4.1.1). The proposed approach is especially useful for non-differentiable dynamical systems. As a result of the brief analysis of the application of the existing algorithms for such systems (see Sec. 5.3), some kind of special treatment of the problem in each given case is required. Even in the most sophisticated and universal Müller method (Müller (1995), Sec. 5.3.3), the function guiding the system through the discontinuity has to be individually defined. On the other hand, the proposed approach is based on a direct numerical investigation of the CS threshold. Hence, the main advantage of the synchronization-based approach is its "insensitivity" to the character of discontinuity, i.e., it can be realized according to the same numerical procedure for each case of the dynamical system. In the final part of this Chapter, some theoretical schemes for other non-typical applications of the synchronization method are briefly approached, i.e., a technique of the largest RLE estimation is demonstrated and an idea of the effective Lyapunov exponent (ELE) is introduced.

6.1 Theoretical Background for the Estimation Procedure

The synchronization-based method of the LLE estimation can be realized both for continuous- and discrete-time systems.

6.1.1 *Continuous-time systems*

The theoretical basis for the LLE estimation procedure is provided with theorem 3.1, in particular with the condition for the CS of two identical oscillators given by inequality (3.17). However, in order to carry out the estimation procedure, a double oscillator system with a unidirectional, uniformly *diagonal coupling* has to be constructed. Such a system is described as follows:

$$\dot{\mathbf{x}} = \mathbf{f}(\mathbf{x}), \tag{6.1a}$$

$$\dot{\mathbf{y}} = \mathbf{f}(\mathbf{y}) + d\mathbf{I}_k(\mathbf{x} - \mathbf{y}), \tag{6.1b}$$

where $\mathbf{x}$, $\mathbf{y} \in \Re^k$ represent the master (reference trajectory) and slave (disturbed trajectory) systems, respectively, $\mathbf{I}_k$ is an $k \times k$ unit matrix, and $d \in \Re$ is a coefficient of coupling strength.

After the introduction of a new variable $\mathbf{z} = \mathbf{y} - \mathbf{x}$, which represents the synchronization error between both oscillators, and the subtraction of Eq. (6.1a) from Eq. (6.1b), the ODE describing time evolution of $\mathbf{z}$ takes the following form:

$$\dot{\mathbf{z}} = \mathbf{f}(\mathbf{x} + \mathbf{z}) - \mathbf{f}(\mathbf{x}) - d\mathbf{I}_k \mathbf{z}. \tag{6.2}$$

The variational equation of Eq. (6.2) is:

$$\dot{\zeta} = \left(D\mathbf{f}(\mathbf{x}(t), \mathbf{x}_0) - d\mathbf{I}_k \right)\zeta, \tag{6.3}$$

where $D\mathbf{f}(\mathbf{x}(t), \mathbf{x}_0)$ is the Jacobi matrix of master system (6.1a), which is initialized from the generic initial conditions $\mathbf{x}_0$. On the basis of this matrix, the LEs of system (6.1a) can be calculated according to classical formula (5.10):

$$\lambda_j = \lim_{t \to \infty} \frac{1}{t} \ln|r_j(t)|, \tag{6.4}$$

where $j = 1, 2, \ldots, k$. From Eq. (6.3) the following relation between the eigenvalues $s_j(t)$ of the Jacobi matrix of the linearized *synchronization error* (Eq. (6.3)), defining the transverse stability of the synchronization manifold $\mathbf{x} = \mathbf{y}$, and eigenvalues $r_j(t)$ of the linearized Jacobi matrix $D\mathbf{f}(\mathbf{x}(t), \mathbf{x}_0)$ of the reference system (6.1a), results:

$$s_j(t) = \exp(-d \cdot t) r_j(t). \tag{6.5}$$

On the basis of the eigenvalues $s_j(t)$ (Eq. (6.5)), the TLEs λ^j_T, quantifying the synchronizability of diagonally coupled oscillators, Eqs. (6.1a–b), can be calculated in the following way:

$$\lambda^j_T = \lim_{t \to \infty} \frac{1}{t} \ln |s_j(t)| = \lim_{t \to \infty} \frac{1}{t} \ln |r_j(t)| + \frac{1}{t} \ln \exp(-d \cdot t). \qquad (6.6)$$

Hence,

$$\lambda^j_T = \lambda_j - d . \qquad (6.7)$$

The synchronous state is stable if all the TLEs are negative, so the largest of them has also to fulfill the condition $\lambda^1_T < 0$. Thus, the condition of the complete synchronization of the reference (Eq. (6.1a)) and perturbed (Eq. (6.1b)) systems is provided by the inequality:

$$d > \lambda_1 , \qquad (6.8)$$

where λ_1 is the LLE of the reference system.

The similar condition of the CS was identified earlier by Fujisaka and Yamada (1983a) and Pikovsky (1984). They proved the existence of an analogous linear dependence between the maximum Lyapunov exponent of the coupled, identical chaotic systems, and the vector of the coupling parameter characterizing such a coupling in the case of the negative feedback mechanism.

As has been explained in Sec. 3.2.1, a *diagonal coupling* of the identical systems (6.1a–b) leads to an appearance (for sufficiently small distance z) of two mutually counteracting effects: an exponential divergence or a convergence (it depends on the sign and the magnitude of λ_1 and d) of the reference $\mathbf{x}(t)$ and disturbed $\mathbf{y}(t)$ trajectories. This property of the *diagonal coupling* causes a linear dependence between the LLE and the value of the coupling coefficient for which synchronization appears (inequality (6.8)). Consequently, it can be used to determine the LLE via numerical investigations of the synchronization process. Such a direct approach works equally well in any arbitrary kind of a dynamical system (not only given by continuous ODEs), because it allows us to avoid a problem of defining the Jacobi matrices for some singular points on the system trajectory, e.g., during a transition via discontinuity.

6.1.2 *Maps*

If a *unidirectional, diagonal coupling*, analogous to Eqs. (6.1a–b), is introduced between the two identical maps, i.e., $\mathbf{x}_{n+1}=\mathbf{f}(\mathbf{x}_n)$ and $\mathbf{y}_{n+1}=\mathbf{f}(\mathbf{y}_n)+d(\mathbf{f}(\mathbf{x}_{n+1}) - \mathbf{f}(\mathbf{y}_{n+1}))$, then the LLE of the master map can be detected in the numerical investigation of synchronization according to the condition for the CS, i.e., $1 - \exp(-\lambda_1) < d < 1 + \exp(-\lambda_1)$, resulting from Eq. (4.5). However, this condition does not give a direct reading of λ_1, like it is in the case of inequality (6.8). Therefore, another scheme of the *master-slave coupling*, which ensures the CS condition in the form given by inequality (6.8), is provided below.

Let us consider a pair of identical and mutually independent maps:

$$\mathbf{x}_{n+1} = \mathbf{f}(\mathbf{x}_n), \tag{6.9a}$$

$$\mathbf{y}_{n+1} = \mathbf{f}(\mathbf{y}_n), \tag{6.9b}$$

which describe an evolution of the phase points $\mathbf{x}_n$, $\mathbf{y}_n \in \mathfrak{R}^k$ on the identical chaotic attractors $\mathcal{M}(\lambda_1 > 0)$. Let us introduce a new variable:

$$\mathbf{z}_n = \mathbf{x}_n - \mathbf{y}_n, \tag{6.10}$$

which can be treated as a disturbance of the state of CS. Using this variable, a *master–slave coupling* between subsystems (6.9a) and (6.9b) can be introduced as:

$$\mathbf{x}_{n+1} = \mathbf{f}(\mathbf{y}_n + \mathbf{z}_n), \tag{6.11a}$$

$$\mathbf{y}_{n+1} = \mathbf{f}(\mathbf{y}_n), \tag{6.11b}$$

$$\mathbf{z}_{n+1} = \exp(-d)[\mathbf{f}(\mathbf{y}_n + \mathbf{z}_n) - \mathbf{f}(\mathbf{y}_n)], \tag{6.11c}$$

where $d \in \mathfrak{R}$ represents the coupling coefficient.

The illustration of the system given in the form of Eqs. (6.11a) and (6.11b) nicely presents that Eq. (6.11a) defines the disturbed orbit depending on the reference trajectory of subsystem (6.11b). It is obvious that the CS of both the maps takes place when the phase distance (*synchronization error*) between them, defined by Eq. (6.11c), reaches the zero value. The condition is fulfilled when all LEs of the disturbance λ^z_j, which characterize Eq. (6.11c), are negative. They can be treated as an equivalent of the TLE in continuous-time systems. To determine them, one should define the Jacobi matrix of Eq. (6.11c) according to formula

(5.14). In such a case, the state of the CS is stable ($\lim_{n\to\infty} \mathbf{z}_n = 0$), and the matrix is described by the following equation:

$$\mathbf{J}_n(\mathbf{z}_0) = \exp(-nd)\mathbf{J}_n(\mathbf{y}_0). \tag{6.12}$$

The matrix $\mathbf{J}_n(\mathbf{y}_0)$ in Eq. (6.12) is a Jacobi matrix of the map $\mathbf{f}^n$, which arises by means of the n-times iteration of the mapping $\mathbf{f}$ in Eq. (6.11b). Thus, the $\mathbf{f}^n$ defines the state $\mathbf{y}_n$ by the initial condition $\mathbf{y}_0$. The matrix $\mathbf{J}_n(\mathbf{y}_0)$ is calculated according to the formula of calculating a derivative of the complex function:

$$\mathbf{J}_n(\mathbf{y}_0) = \prod_{i=0}^{n-1} \mathbf{J}(\mathbf{y}_i), \tag{6.13}$$

where $\mathbf{J}(\mathbf{y}_i)$ are Jacobi matrices defined in points $\mathbf{y}_i$. The Lyapunov exponents of the disturbance can be calculated according to:

$$\lambda_i^z = \lim_{n\to\infty} \frac{1}{n} \ln|s_i(n)|, \tag{6.14}$$

where $s_i(n)$ are the eigenvalues of the $\mathbf{J}_n(\mathbf{z}_0)$ matrix equal to:

$$s_i(n) = \exp(-nd)\sigma_i(n), \tag{6.15}$$

where $\sigma_i(n)$ are eigenvalues of the $\mathbf{J}_n(\mathbf{y}_0)$ matrix. Thus, the exponents λ_i^z are in fact the TLEs calculated on the basis of linearized equation:

$$\dot{\zeta} = \mathbf{J}_n(\mathbf{z}_0)\zeta. \tag{6.16}$$

Substituting formula (6.15) into Eq. (6.14), we obtain:

$$\lambda_i^z = \lim_{n\to\infty} \frac{1}{n} \ln|\sigma_i(n)| + \lim_{n\to\infty} \frac{1}{n} \ln \exp(-nd). \tag{6.17}$$

Using Eq. (5.14) in Eq. (6.16), we have:

$$\lambda_i^z = \lambda_i - d, \tag{6.18}$$

where λ_i are the Lyapunov exponents of mapping (6.11b). The state of the CS of subsystems (6.11a) and (6.11b) assures the fulfillment of the inequality $\lambda_i^z < 0$, which can be expressed with the following theorem.

Theorem 6.1 *Let us consider a dynamical system described by differential Eqs. (6.11a–c). If λ_1 is the largest Lyapunov exponent characterizing an evolution on the attractor $\mathcal{M}$ of mapping (6.11b), then fulfillment of inequality (6.8), i.e.,*

$$d > \lambda_1$$

makes the full synchronization of mappings (6.11a) and (6.11b) possible.

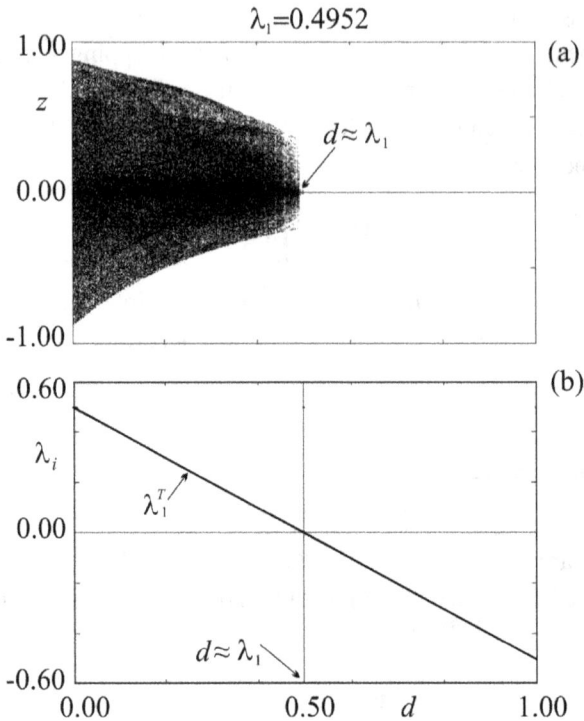

Fig. 6.1. Bifurcational diagrams presenting the disturbance z (a), and, the corresponding LLE (b), as a function of the coupling parameter d, for coupled chaotic, logistic maps — Eq. (2.48); $r=3.90$. The value of λ_1 is calculated using a classical algorithm — in the top part.

The same synchronization condition for maps and flows given by inequality (6.8) shows directly that the form of the equation defining an evolution of disturbance (6.11c) was not arbitrary chosen. This kind of coupling was introduced to damp exponentially the perturbation z when the coupling strength increases over the value of the LLE. Then, the mechanism leading to the CS of mappings (6.11a) and (6.11b) possesses properties identical to the *negative feedback* of the phase flows in the case of an *unidirectional coupling*, i.e., when the relation between the subsystems takes a form of the *master–slave coupling*. The constant component "exp($-d$)" is modelling an effect of the coupling from systems (3.15) and (6.1a–b) in Eqs. (6.11a–c), and its influence as attracting the trajectories $\mathbf{x}_n$ i $\mathbf{y}_n$ in the phase space possesses the same isotropic character as shown in Fig. 3.1b.

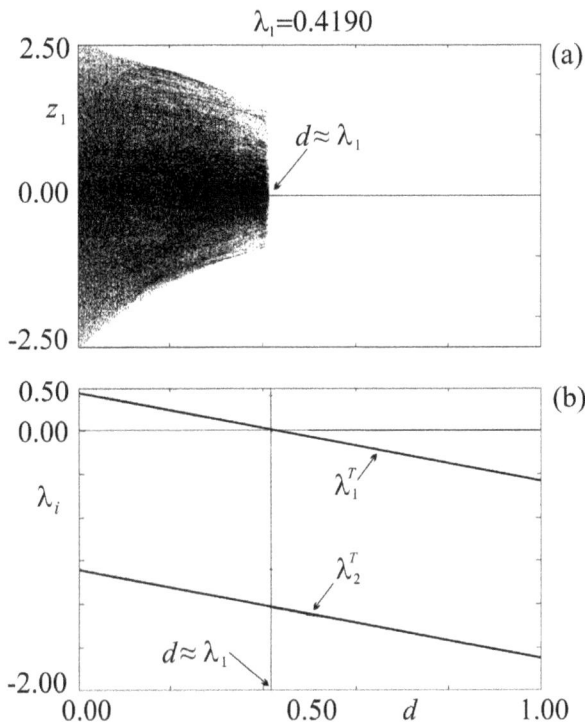

Fig. 6.2. Bifurcational diagrams presenting the disturbance z (a), and, two corresponding LLE (b), as a function of the coupling parameter d, for coupled chaotic, Henon maps — Eq. (2.49); $a=1.40$, $b=0.30$. The value of λ_1 is calculated using a classical algorithm — in the top part.

A confirmation of the analogies between synchronization of the phase flows and the mappings discussed here are identical synchronization conditions (6.8) and (6.18). The Figs. 6.1a–b and 6.2a–b present bifurcational diagrams of the trajectory separation z as a function of the parameter d and traces of the Lyapunov exponents related to them, for two classical discrete mappings — the logistic map:

$$y_{n+1} = ry_n(1-y_n), \tag{6.19}$$

and the Henon map:

$$y_1^* = 1 - ay_1^2 + y_2,$$
$$y_2^* = by_1. \tag{6.20}$$

After substituting the above equations to formulas (6.11a–c), we obtain the following expanded sets of differential equations:

$$x_{n+1} = r(y_n + z_n)[(1-(y_n + z_n))],$$
$$y_{n+1} = ry_n(1-y_n),$$
$$z_{n+1} = r(z_n - z_n^2 - 2y_n z_n)\exp(-d), \tag{6.21}$$

for the logistic map, and,

$$x_1^* = 1 - a(y_1 + z_1)^2 + (y_2 + z_2),$$
$$x_2^* = b(y_1 + z_1),$$
$$y_1^* = 1 - ay_1^2 + y_2,$$
$$y_2^* = by_1,$$
$$z_1^* = [z_2 - az_1(z_1 + 2y_1)]\exp(-d),$$
$$z_2^* = bz_1 \exp(-d), \tag{6.22}$$

for the Henon mapping.

For chosen values of the parameters a, b, r, the attractors of the discussed mappings are chaotic. In Figs. 6.1a and 6.2a one can observe that a possibility of the synchronization of subsystems (6.11a) and (6.11b) is limited by condition (6.18), similar to condition (6.8) for the phase flows. This analogy confirms that the coupling defined by Eqs. (6.11a–c) precisely models the phenomenon of the CS of the phase flows with a *unidirectional, negative feedback coupling*.

6.2 Numerical Implementation

Inequalities (6.8) and (6.18) allow us to estimate the LLE for any type of the dynamical system. In fact, these inequalities state that the smallest value of the coupling coefficient d, for which synchronization occurs, is approximately equal to λ_1.

6.2.1 Variants of the method

In order to apply the synchronization method in numerical simulations, it is necessary to build an augmented system according to Eqs. (6.1a–b) or (6.11a–c). The next step is a numerical search for the synchronous value of d approaching the LLE of the investigated system. The estimation process can be conducted in three variants:

1. from an integration of augmented systems (Eqs. (6.1a–b)) — for continuous-time systems (*variant* 1 of the method in further considerations),
2. from iterations of coupled maps (Eqs. (6.11b) and (6.11c)) — for discrete-time systems given by the known difference equations (*variant* 2),
3. from iterations of coupled maps (Eqs. (6.11a–c)), where maps are reconstructed from the flow or they are not explicitly defined (*variant* 3).

6.2.2 Procedures improving the effectiveness of the method

The simplest way to evaluate the smallest synchronous value of the coupling d is to construct a bifurcation diagram of the *synchronization error z* versus d. The magnitude of the *synchronization error* is represented by Eq. (6.11c) in *variants* 2 and 3 described above or it can be computed according to the following formula:

$$z = \|\mathbf{z}\| = \sqrt{\sum\nolimits_{i=1}^{k}(x_i - y_i)^2} \, , \tag{6.23}$$

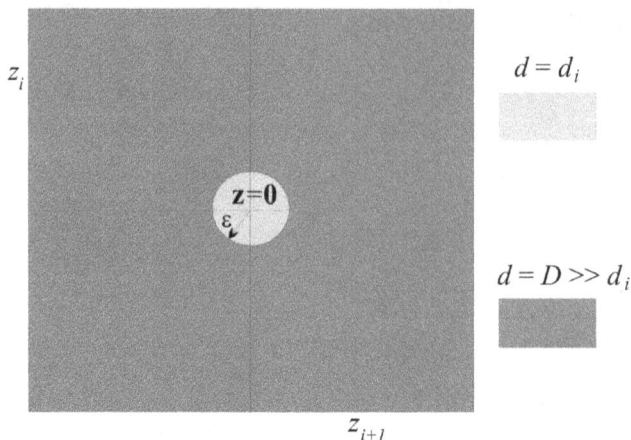

Fig. 6.3. Visualization of the *elastic coupling* idea.

in *variant* 1 of the method. Then, the LLE can be estimated as a value of d where z approaches zero. However, such a way for the LLE evaluation is usually very time-consuming. Therefore, for the calculations presented in this paper, we have applied a method of fast searching the synchronous value of d. This method exploits procedures improving its effectiveness, which are shortly described below.

(1) *Elastic coupling*, i.e., a dependence of the parameter d on the synchronization error z (Eq. (6.23)) defined by the formula:

$$d = \begin{cases} d_i & \text{for} \quad z \le \varepsilon, \\ D & \text{for} \quad z > \varepsilon, \end{cases} \tag{6.24}$$

where D is a strong coupling parameter, much larger than its currently tested value d_i ($D>>d_i$), and ε is a small parameter. Such an *elastic coupling* allows us to reduce significantly a disadvantageous effect of a long-time transient motion before the appearance of the CS, because the trajectory of the slave system is forced to evolve in the neighborhood of the reference trajectory of the master system, which accelerates the CS process. To achieve faster the synchronization value d, the phase space of the researched system has been divided into two regions (see Fig. 6.3). The first one is a direct neighborhood of the synchronization subspace $\mathbf{x} = \mathbf{y}$. This region is bounded by a coefficient ε which is a radius of the k-dimensional sphere with a center in $\mathbf{z} = 0$ (light gray). The second region is the remaining part of the phase space (dark gray). Thus, the main idea of this method is a sudden jump of the coupling coefficient when the *synchronization error* crosses the boundary ε value. Inside the sphere $z \leq \varepsilon$ (Fig. 6.3), the coupling parameter has the investigated d_i value. Beyond the ICS region ($z > \varepsilon$), d assumes its strong rate D (Eq. (6.24)).

(2) *Imperfect complete synchronization* (ICS) defined by inequality (1.4), where the small parameter ε plays a role of the ICS threshold (the same as in Eq. (6.24)). An application of the ICS also allows us to reduce the estimation time, because in practice it is enough to confirm that the synchronization state is asymptotically stable for the currently investigated value d_i. For this purpose, for each iteration d_i, a constant estimation period T_e (or a limited number of the map steps n_e) and the transient period T_t (n_t) has to be assumed (Fig. 6.4). If during the testing period, the ICS condition resulting from inequality (1.4), i.e.,

$$z \leq \varepsilon \quad \text{for} \quad T_t < t \leq T_e \quad \text{or} \quad n_t < n \leq n_e, \tag{6.25}$$

is fulfilled all the time (for all iterations), then the synchronization is stable (so the CS is possible) for the tested d_i (see Fig. 6.4a). On the other hand, if even in one moment of the testing time T_t (n_t), the distance z crosses the boundary value ε ($z > \varepsilon$ for $T_t \geq t > T_e$ or $n_t \geq n > n_e$), then the synchronization is recognized to be unstable (see Fig. 6.4b).

(3) *Iterative bisection* of the tested range of the coupling parameter $\Delta d_i = d_{1(i)} - d_{2(i)}$, where $d_{1(i)}$, $d_{2(i)}$ are the left and right end of the d-range after the i-th iteration of the bisection ($i = 0, 1, 2, \ldots, i_{max}$), correspondingly. The currently tested coupling parameter in the i-th iteration depends on an occurrence of the synchronization ($d_i = d_{2(i-1)} - \Delta d_{i-1}/2$) or a lack of it ($d_i = d_{1(i-1)} + \Delta d_{i-1}/2$) until the range Δd_i is less than the assumed precision of estimation δ. Thus, the number of iterations i_{max} required to estimate the LLE with the precision δ can be calculated as follows:

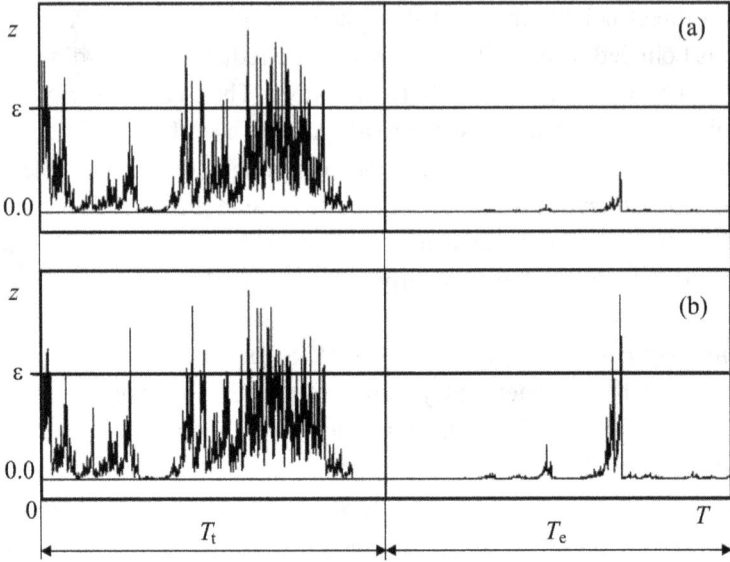

Fig. 6.4. Exemplary time diagrams of the *synchronization error* (Eq. (6.23)) illustrating a practical criterion for the ICS (a) or its lack (b) in accordance with the assumed formula (6.25).

$$i_{max} = \text{int}\left[\log_2\left(\frac{\Delta d_0}{\delta}\right)\right] + 1, \tag{6.26}$$

where int[...] is a function returning the integer number from the value in the quadratic brackets, and $\Delta d_0 = d_{1(0)} - d_{2(0)}$ (see Fig. 6.5) represents the initial d-range. A quick estimation of the LLE with this method is graphically illustrated in Fig. 6.5. This example shows that for an estimation of the value of λ_1 (for the classical Henon map, Eq. (6.20)) with an accuracy $\delta d = 0.001$, only 12 iterations steps of the bisection procedure were sufficient ($i_{max} = 12$).

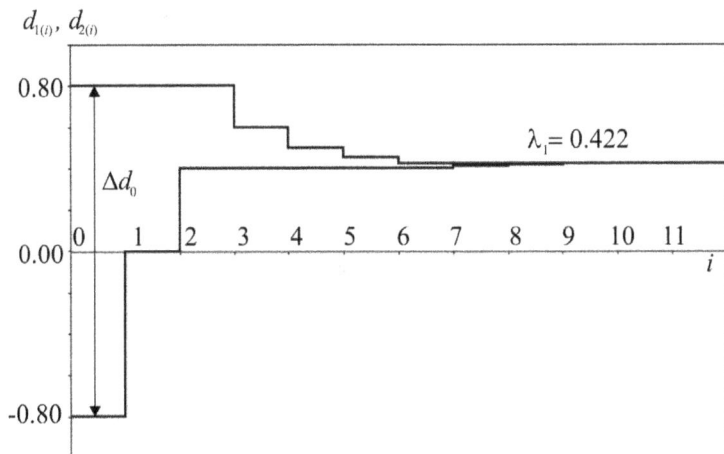

Fig. 6.5. Trace of the estimation of the LLE for the Henon system (Eq. (6.20)) by the bisection method (scheme in Fig. 6.6) conducted on the example of the coupled Henon maps (Eq. (6.22)); values of the system parameters: a=1.40, b=0.30; values of the numerical procedure parameters: ε = 0.01, D = 1.00, n_t = 5000, n_e= 5000, $d_{1(0)}$ = –0.80, $d_{2(0)}$ = 0.80, δd = 0.001.

Table 6.1. Suggested optimal ranges of the constants applied in the estimation process.

Constant	Range	Comment
ε	$10^{-2}\div10^{-4}\times|\mathcal{A}|$	$|\mathcal{A}|$ — size of the attractor ($|\mathcal{A}| = \sup(z)$)
D	$2\div10\times\lambda_1$	Optimal value of D assures the CS of the investigated subsystems (at the constant $d=D$) in time of less than 100 orbital cycles
n_t	$300\div5000$	Such a value has been used for the maps generated in *variants 2* or *3*. In the case of *variant 1*, it remains for a number of the orbital cycles. Such a cycle can be a period of excitation $T=2\pi/\eta$ for the Duffing system.
n_e	$500\div10000$	
$d_{1(0)}$	$<\lambda_1$	Arbitrary value (possibly large) which does not allow for synchronization in the complete range of a bifurcation parameter
$d_{2(0)}$	$>\lambda_1$	Arbitrary value (possibly small) which allows for synchronization in the complete range of a bifurcation parameter
δd	$10^{-2}\div10^{-3}$	Lower values increase the time of estimation without any practical improvement in accuracy

INPUT:
initial conditions: $\mathbf{x}_0, \mathbf{y}_0$;
parameters of numerical procedure:
$\varepsilon, D, d_{1(0)}, d_{2(0)}, \delta, n_t, n_e$

$i=1, d_{1(i-1)}=d_{1(0)}, d_{2(i-1)}=d_{2(0)}$
$\Delta d_{i-1}=d_{2(0)}-d_{1(0)}$

$d_i=d_{2(i)}=d_{2(i-1)}-\dfrac{\Delta d_{i-1}}{2}$
$d_{1(i)}=d_{1(i-1)}$

$n=0, \mathbf{x}_0 \neq \mathbf{y}_0$

$|\mathbf{x}_n - \mathbf{y}_n| < \varepsilon$ YES   NO

$n > n_e$ YES   NO

$n > n_t$ YES   NO

$d = d_i$

$d = D$

$\mathbf{x}_n \rightarrow \mathbf{x}_{n+1}$ $\mathbf{y}_n \rightarrow \mathbf{y}_{n+1}$

$n = n+1$

$i = i+1$

$|\mathbf{x}_n - \mathbf{y}_n| = 0$ NO   YES

$\Delta d_i = d_{2(i)} - d_{1(i)}$

$d_{1(i)}=d_{1(i-1)}+\dfrac{\Delta d_{i-1}}{2}$
$d_{2(i)}=d_{2(i-1)}$

$\Delta d_i < \delta$ NO   YES

OUTPUT:
$\lambda_1^R = \dfrac{d_{1(i)}+d_{2(i)}}{2}$

next value of bifurcational parameter

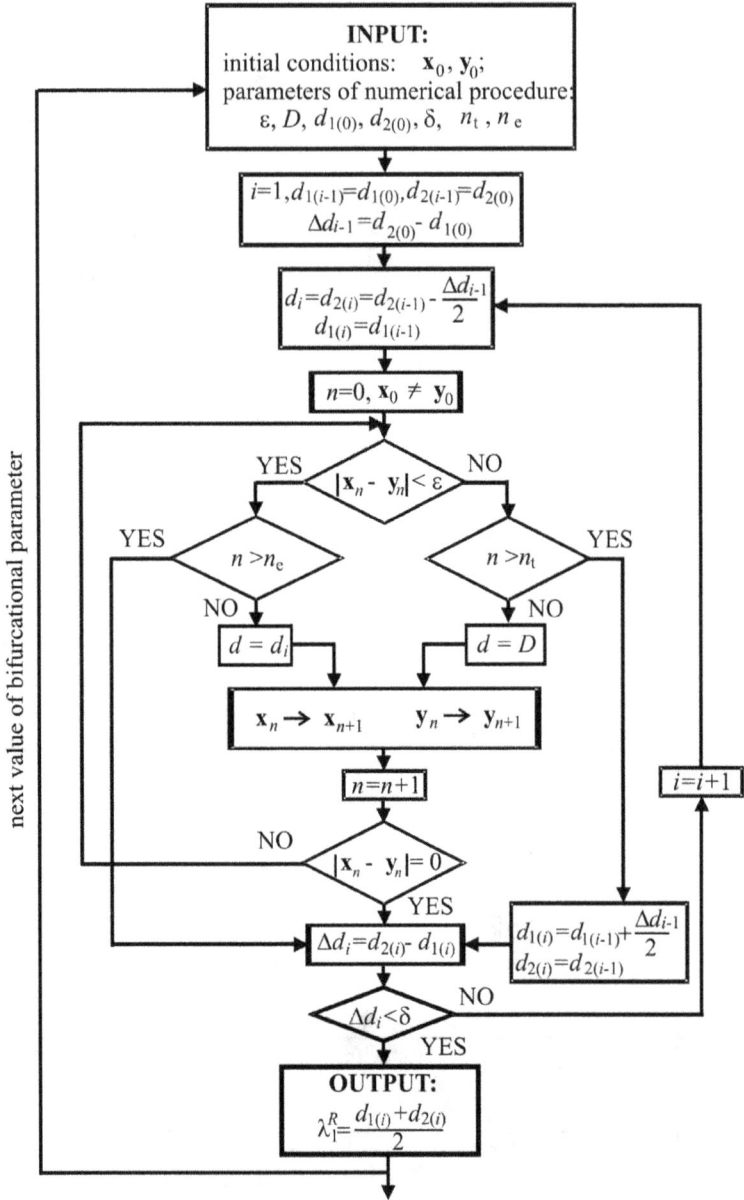

Fig. 6.6. Block scheme of the numerical estimation the largest RLE including auxiliary procedures: (1) *iterative bisection*, (2) *elastic coupling*, and (3) *imperfect complete synchronization*.

6.3 Estimation of the Largest RLE

Examples of the RLEs estimated with the proposed synchronization based approach are presented in Chapter 4 (Subsection 4.3.2). In order to adapt the estimation procedure to determination of the RLEs of externally driven oscillators (Eqs. (4.60a–b)), a double response system with a common drive and an additional *diagonal coupling* between the response oscillators has to be constructed. Such a system is described as follows:

$$\dot{\mathbf{e}} = \mathbf{g}(\mathbf{e}), \tag{6.27a}$$

$$\dot{\mathbf{x}} = \mathbf{f}(\mathbf{x}) + q\mathbf{h}(\mathbf{e}), \tag{6.27b}$$

$$\dot{\mathbf{y}} = \mathbf{f}(\mathbf{y}) + q\mathbf{h}(\mathbf{e}) + d(\mathbf{x} - \mathbf{y}), \tag{6.27c}$$

where $\mathbf{x}$ and $\mathbf{y}$ represent the *master response* (MRS) and auxiliary, identical *slave response* (SRS) *systems*, respectively, and d is a coefficient of the coupling strength.

By analogy, an augmented system for estimation of the RLE of externally driven maps (Eqs. (4.61a–b)) is as follows:

$$\mathbf{e}_{n+1} = \mathbf{g}(\mathbf{e}_n), \tag{6.28a}$$

$$\mathbf{x}_{n+1} = \mathbf{f}(\mathbf{x}_n) + q\mathbf{h}(\mathbf{e}_n), \tag{6.28b}$$

$$\mathbf{y}_{n+1} = \mathbf{f}(\mathbf{x}_n + \mathbf{z}_n) + q\mathbf{h}(\mathbf{e}_n), \tag{6.28c}$$

$$\mathbf{z}_{n+1} = [\mathbf{f}(\mathbf{x}_n + \mathbf{z}_n) - \mathbf{f}(\mathbf{x}_n)]\exp(-d), \tag{6.28d}$$

where Eqs. (6.28b) and (6.28c) describe an evolution of the MRS and the SRS, respectively, whereas Eq. (6.28d) governs the dynamics of the *synchronization error* between response systems under control of the coupling strength d.

The process of estimation of the largest RLE is realized via numerical investigation of the CS of response oscillators (6.27b–c) for continuous-time systems, or (6.28b–c) for maps, according to the numerical procedures presented in Sec. 6.2.

For the needs of estimation of the largest RLE of Duffing oscillators (Eq. 4.62) driven by the signal $e_1(t)$ from the Lorenz system (Eq. (4.63)) or a random *non-deterministic* signal produced from Eqs. (4.64a–b), presented in Sec. 4.3.2., a double slave system with an auxiliary SRS has to be constructed in order to perform *variant* 1 of the estimation procedure according to Eqs. (6.27a–c):

$$\dot{x}_1 = x_2,$$
$$\dot{x}_2 = -\alpha x_1^3 - hx_2 + q[e(t) - x_1],$$

(6.29a)

$$\dot{y}_1 = y_2 + d(x_1 - y_1),$$
$$\dot{y}_2 = -\alpha y_1^3 - hy_2 + q[e(t) - y_1] + d(x_2 - y_2),$$

(6.29b)

where the uniform diagonal *master–slave coupling* between Eqs. (6.29a) and (6.29b) is realized, according to the scheme shown in Fig. 6.7.

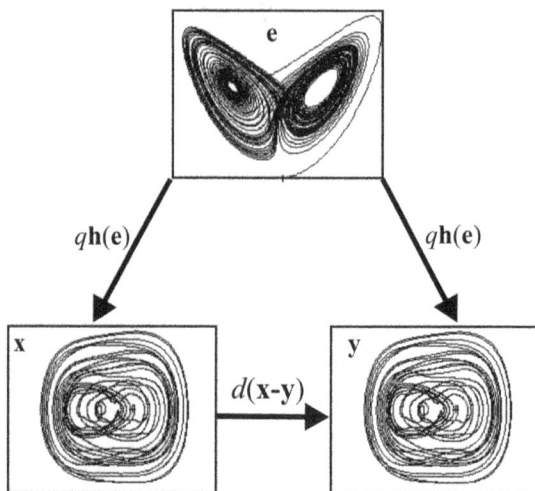

Fig. 6.7. Example of an additional *master–slave coupling* of two response systems: two Duffing oscillators (MRS and SRS, see Eq. (4.62)) driven by the chaotic Lorenz system (Eq. (4.63)), introduced in order to realize the synchronization based estimation procedure.

On the other hand, in order to realize the proposed method of estimation of the largest RLE for Henon maps (Eq. (4.65)) as response systems driven by a logistic mapping (Eq. (4.66)) or a random signal (Eq. (4.67)), the auxiliary SRS Henon system:

$$y_1^* = 1 - a(x_1 + z_1)^2 + (x_2 + z_2) + qe,$$
$$y_2^* = b(x_1 + z_1),$$
(6.30)

and the *synchronization error* system:

$$z_1^* = [z_2 - az_1(z_1 + 2x_1)]\exp(-d),$$
$$z_2^* = bz_1 \exp(-d),$$
(6.31)

have to be taken under consideration in the numerical simulations. The largest RLE of the driven Henon system (Eq. (4.65)) can be estimated by means of *variant* 2 of the method in both the cases of the drive (*deterministic* and *non-deterministic*) due to the known difference equations describing the system. In this variant, Eq. (6.30) can be neglected. Obviously, we can use here also *variant* 3 of the method, which is more general than *variant* 2 and allows us to estimate the RLE in cases when an equation of the map is not clearly given, e.g., Poincaré cross-sections which are reconstructed from the continuous-time system.

6.4 Effective Lyapunov Exponents

After a small modification, the synchronization method can be effectively applied for the systems with noise or time-varying parameters. This technique can be treated as a slight development of its existing version, which is based on the properties of the CS via a *diagonal coupling* of two identical systems. The method has been modified by introducing a small mismatch between them, which models an influence of the noise on the system dynamics.

The dynamical nature of such disturbed systems is slightly different in comparison to typical, unperturbed oscillators. Generally, the noise can be treated as a relatively small but unexpected disturbance of the input/output signal or the system parameters, which is of unknown source and character. Thus, the classical algorithmic methods for calculating the LEs cannot be used here because such a system is non-differentiable. Normally, in a periodic system, the convergence of the nearby trajectories leading to the CS should be observed. However, after introducing the noise there appears a divergence of close orbits due to the permanent disturbance caused by the noise, while the general dynamics of the systems remains still regular. In order to model the noise, a mismatch between the reference and disturbed systems is introduced by adding the noise component to the slave system. Then, Eq. (6.1a–b) assumes the form:

$$\dot{\mathbf{x}} = \mathbf{f}(\mathbf{x})$$
$$\dot{\mathbf{y}} = \mathbf{f}(\mathbf{y}) + \Delta(\mathbf{y}) + d\mathbf{I}_k(\mathbf{x} - \mathbf{y}) \tag{6.32}$$

where $\Delta(\mathbf{y}) \in \mathfrak{R}^k$ represents a *mismatch vector*. Applying the proposed method for slightly different, when uncoupled, systems (6.32), we can estimate a magnitude of the divergence of both trajectories in the presence of noise, which is quantified with a coupling parameter d providing their ICS, i.e., approaching the practical value of the LLE. Such an *effective Lyapunov exponent* (ELE) can be treated as a measure of the ICS robustness between the reference system and its replica disturbed by the noise. The most decisive factor for the ELE evaluation is the ICS threshold ε assumed in the numerical investigations. Hence, the ELE can be defined as follows:

Definition 6.1 *The number called the effective Lyapunov exponent λ_E is equal to the minimum strength of the diagonal coupling coefficient d linking the reference and disturbed (with noise) systems (Eqs. (6.32)), which is required to maintain the synchronization error z between them in the specified ε-range.*

The ELE has an analogical practical sense like the LLE, because it is also based on the definition of stability by Lyapunov (Definition 5.1). Therefore, it has been named similarly, although the above informal definition of the ELE is not related directly to the basic definition of Lyapunov exponents (Oseledec (1968)). A factor which plays a crucial role in estimating the ELE is the boundary parameter ε (Eq. (23)) defining the condition of the ICS. A proposal how to evaluate its value is presented in Sec. 7.4.

The concept of the ELE can be also applied to quantify the stability of the effective ICS in externally driven oscillators (see Sec. 4.3.2). Hence, an idea of the *effective response Lyapunov exponent* (ERLE) has been introduced. The ERLE can be determined with the same numerical procedure as for the RLE estimation, i.e., on the basis of Eqs. (6.27a–c) or (6.28a–d), but with the parameter mismatch between the MRS and the SRS. It can be performed including a component representing the mismatch into the SRS. Then, Eqs. (6.27c) or (6.28c) assume the forms:

$$\dot{\mathbf{y}} = \mathbf{f}(\mathbf{y}) + \Delta(\mathbf{y}) + q\mathbf{h}(\mathbf{e}) + d\mathbf{I}_m(\mathbf{x} - \mathbf{y}) \tag{6.33}$$

and

$$\mathbf{y}_{n+1} = \mathbf{f}(\mathbf{x}_n + \mathbf{z}_n) + \Delta(\mathbf{x}_n + \mathbf{z}_n) + q\mathbf{h}(\mathbf{e}_n) \tag{6.34}$$

respectively, where $\Delta(\mathbf{y})$, $\Delta(\mathbf{x}_n + \mathbf{z}_n) \in \mathfrak{R}^k$ represent a *mismatch vector*. The inclusion of the mismatch changes also the *synchronization error* map (Eq. (6.28d)):

$$\mathbf{z}_{n+1} = [\mathbf{f}(\mathbf{x}_n + \mathbf{z}_n) + \Delta(\mathbf{x}_n + \mathbf{z}_n) - \mathbf{f}(\mathbf{x}_n)]\exp(-d) . \tag{6.35}$$

Chapter 7

Applications of the Synchronization Method for the LLE Estimation

In this last Chapter, a few examples of the LLE estimation with the numerical algorithm, which has been described in the previous Chapter, are demonstrated. The subject of the analysis are numerical models of selected dynamic systems, both maps and phase flows. For a more detailed presentation of advantages of the proposed method, this review of its applications is mainly composed of examples of dynamical systems with discontinuities or with time delay, for which the use of any well-known method of calculating the Lyapunov exponents is difficult or not straightforward (see Chapter 5).

The results of estimation are given in the form of individual values received for specific sets of parameters of the tested system, as bifurcation diagrams are functions of selected coefficients, as well as forms of space-images or colored maps, illustrating the LLE in the two-dimensional space of the system parameters. Furthermore, it should be noted that the value received by λ_1 obtained with *variants* 1 and 2 characterize directly the phase flow or the map given by difference equations. By contrast, in the case of the use of *variant* 3, the values given in Figs. as λ_1 characterize the phase flow from which the map was reconstructed, i.e., they are already calculated according to formula (5.31). On the other hand, if the presented Lyapunov exponents are directly related to the map, they are marked differently, e.g., λ_{1P} for the Poincaré map. The presented examples are divided into four groups, namely:

1. discrete maps described with known difference equations,
2. mechanical systems with one or more degrees of freedom with discontinuities as impacts or dry friction,
3. dynamical systems (maps and flows) with time delay,
4. instances of the ELE and ERLE estimation.

7.1 Discrete-time Systems of Known Difference Equations

The classical examples of discrete maps described with difference equations, already presented in Sec. 6.1, are logistic and Henon maps (Eqs. (6.19) and (6.20), respectively). The estimation was realized numerically according to *variant* 2 of the method and based on mathematical models of the augmented systems given by Eq. (6.21) for the

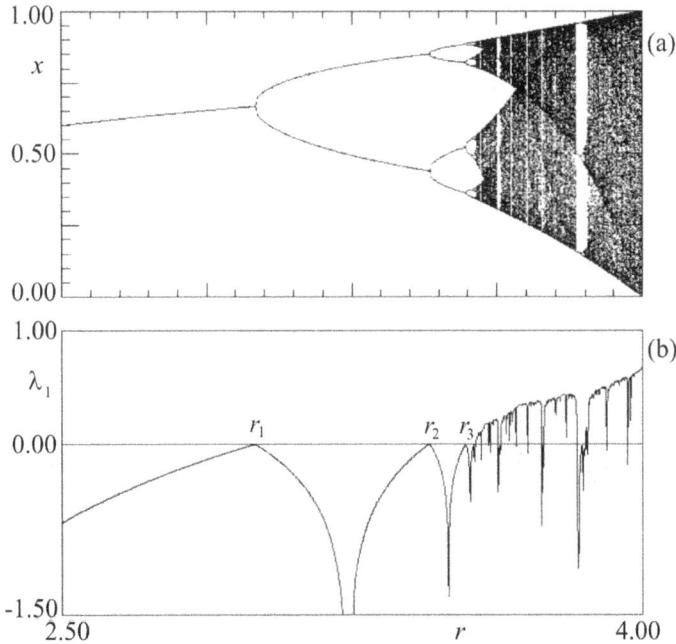

Fig. 7.1. Bifurcation diagram of the logistic map (Eq. (6.19)) as a function of the parameter r (a), and the corresponding LLE (b), estimated by the synchronization method; numerical procedure parameters: $\varepsilon=0.001$, $D=2.00$, $n_t=5000$, $n_e=5000$, $d_{1(0)}=-1.00$, $d_{2(0)}=1.00$, $\delta=0.001$.

logistic map, and Eq. (6.22) for the Henon map. It should be mentioned here that *variant* 3 of the method is also efficient in these cases.

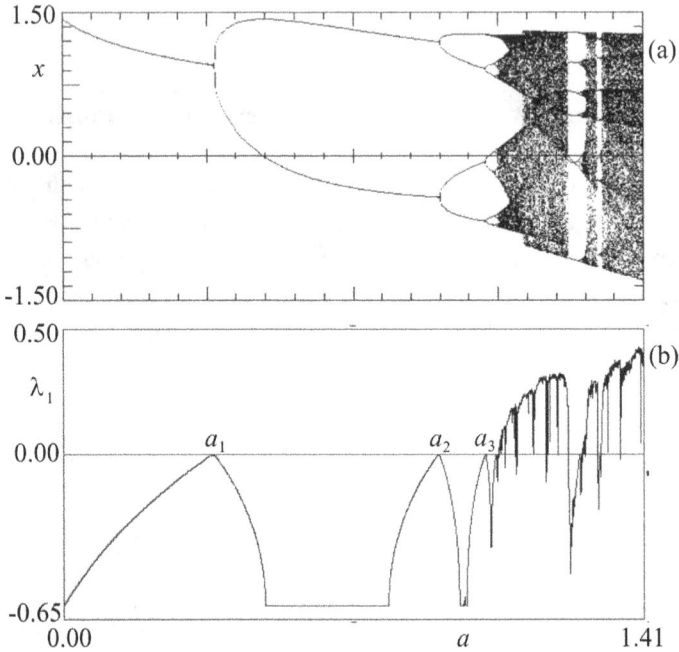

Fig. 7.2. Bifurcational diagram of the Henon map (Eq. (6.20)) was a function of the parameter a (a), and the related trace of the LLE (b), estimated by the synchronization method for $b=0.30$; numerical procedure parameters: $\varepsilon=0.001$, $D=2.00$, $n_t=5000$, $n_e=5000$, $d_{1(0)}=-1.00$, $d_{2(0)}=1.00$, $\delta d=0.001$.

In Figs. 6.1a and 6.2a estimation examples of single values of the LLE with bifurcation diagrams of the *disturbance (synchronization error)* z value versus the d coefficient are depicted. On the other hand, Figs. 7.1b and 7.2b present bifurcational diagrams of the LLE values for systems (6.19) and (6.20) as functions of chosen parameters, corresponding to the bifurcational diagrams 7.1a and 7.2a, respectively. It can be seen that the obtained values of the λ_1 properly inform about a character of the motion of the analyzed systems (they are negative in ranges where the motion appears to be periodic and positive where the chaotic motion is realized). Also the bifurcational values r_1, r_2, r_3 (Fig. 7.1b) or a_1, a_2, a_3 (Fig. 7.2b) of control parameters are precisely displayed. They are related to the bifurcation of period-doubling — the λ_1 reaches there values close to zero.

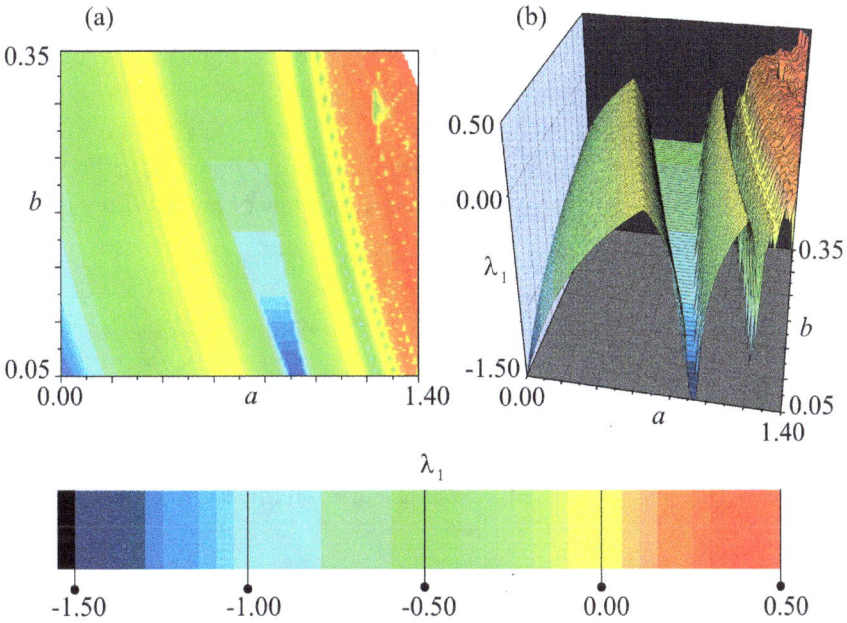

Fig. 7.3. LLE of the Henon map (Eqs. (2.49)) estimated with *variant* 2 of the synchronization method, as a coloured map (a), and a 3-D perspective view (b), as a function of the two parameters a and b. Numerical procedure parameters: $\varepsilon = 0.001$, $D = 2.00$, $n_t = 5000$, $n_e = 5000$, $d_{1(0)} = -1.00$, $d_{2(0)} = 1.00$, $\delta d = 0.001$.

The effectiveness and relatively "high speed" of the estimation method allows for analysis of the dynamics in a wider range of the system parameters. Figures 7.3a and 7.3b illustrate results of the estimation of the LLE as a function of both the parameters a and b of the Henon map. The colors are related to the value of λ_1, which allows for the evaluation of the motion character and the identification of the type of bifurcations taking place in the two-dimensional space of parameters of the system under investigation.

7.2 Mechanical Systems with Discontinuities

A test of the estimation method for mechanical systems with discontinuities is presented below by means of four examples.

7.2.1 1-DoF impact oscillator

A model of the analyzed oscillator is shown in Fig. 7.4. A typical, mechanical 1-DoF system with linear spring and viscous damper characteristics, and external, harmonic excitation realizes the motion limited with a single bumper. Dynamics of the oscillator is described with the 2nd order ODE (Eq. (7.1a)) and the Newton's law of impact (Eq. (7.1b)):

$$x < \delta_0 \quad \Rightarrow \quad m_0\ddot{x} + c_0\dot{x} + k_0 x = P_0 \cos(\Omega t), \qquad (7.1a)$$

$$x \geq \delta_0 \quad \Rightarrow \quad \dot{x}^+ = -R\dot{x}^-, \qquad (7.1b)$$

where m_0 is an oscillator mass, k_0 — spring stiffness, c_0 — viscous damping coefficient, P_0 — exciting force amplitude, Ω — exciting frequency, δ_0 — bumper position, R — restitution coefficient. The values of $\dot{x}^+$ and $\dot{x}^-$ stand for the velocities just before and just after an impact, respectively.

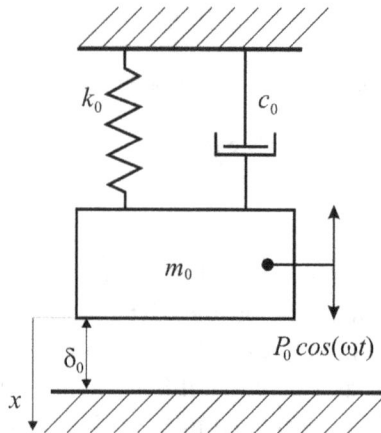

Fig. 7.4. 1-DoF impact oscillator.

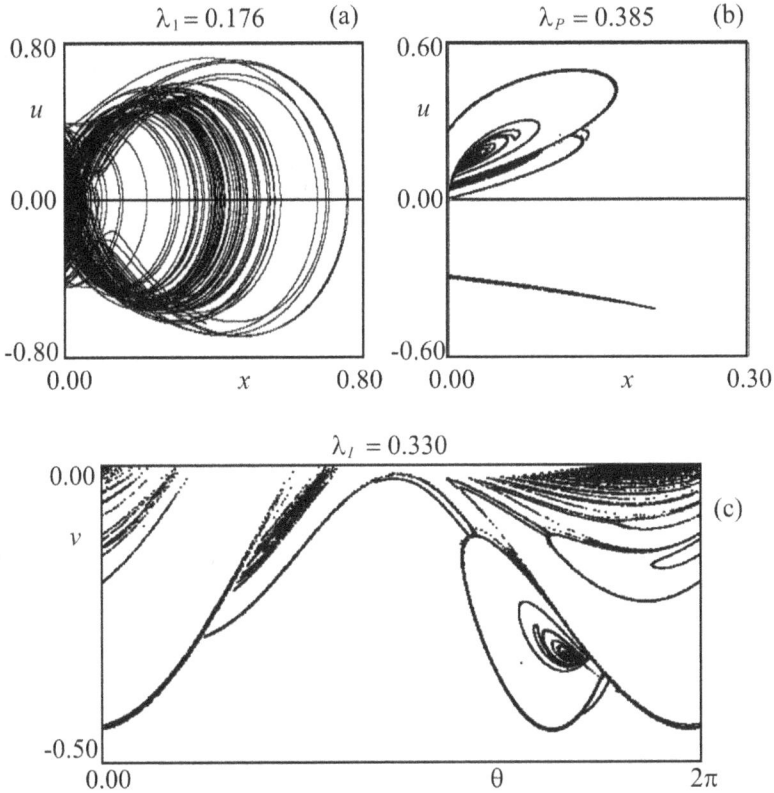

Fig. 7.5. Phase portrait (a), Poincaré map (b), and an impact map (c) of the mechanical oscillator with impacts (Eqs. (7.1a–b)) and the corresponding LLE estimated by the synchronization method (top part); $\eta = 3.150$, $h = 0.10$, $p = 1.00$, $R = 0.90$, $\delta_0 = 0.00$; numerical procedure parameters: $\varepsilon = 0.001$, $D = 3.00$, $n_t = 2000$, $n_e = 2000$, $d_{1(0)} = -0.80$, $d_{2(0)} = 0.80$, $\delta = 0.005$.

Dividing Eq. (7.1a) by k_0, assuming the static deflection $x_{st} = P_0/k_0$, and introducing the normalized time $\tau = \omega t$, where $\omega^2 = k_0/m_0$, we obtain Eq. (7.1a) in the following form of 1st order differential equations:

$$\dot{y}_1 = y_2,$$
$$\dot{y}_2 = p\cos(\eta t) - y_1 - 2hy_2,$$

(7.2)

where: $h = \dfrac{c_0}{\sqrt{k_1/m_1}}$, $\eta = \dfrac{\Omega}{\omega}$, $p = P_0/k_0 x_{st}$, $\dot{y} = \dfrac{1}{\omega}\dfrac{dx}{dt}$ $\ddot{y} = \dfrac{1}{\omega^2}\dfrac{d^2x}{dt^2}$.

The LLE of system (7.1a–b) can be estimated according to *variants* 1 and 3. After substitution of system (7.2), with impact condition (7.1b), into system (6.1a–b), the augmented system for estimating the LLE can be written in the following form:

$$x, y < \delta_0 \quad \Rightarrow \quad \begin{cases} \dot{x}_1 = x_2 + d(y_1 - x_1), \\ \dot{x}_2 = p\cos(\eta t) - x_1 - hx_2 + d(y_2 - x_2), \\ \dot{y}_1 = y_2, \\ \dot{y}_2 = p\cos(\eta t) - y_1 - hy_2, \end{cases}$$

(7.3)

$$x, y \geq \delta_0 \quad \Rightarrow \quad \begin{cases} x_2^+ = -Rx_2^-, \\ y_2^+ = -Ry_2^-, \end{cases}$$

which allows for conducting the process of estimation of the LLE using *variant* 1 of the method.

   Variant 3 can be used both for the Poincaré map (see Fig. 7.5b), and the impact map (Fig. 7.5c), respectively. These mappings can be described with general equations as follows:

$$\begin{aligned} v_{n+1} &= f(v_n, \xi_n), \\ \xi_{n+1} &= g(v_n, \xi_n), \end{aligned}$$

(7.4)

where v and ξ represent the velocity u and the position x on the Poincaré map after each period of the harmonic excitation force, or, the velocity v and the phase of the external excitation θ at the moment of contact on the impact map, respectively. The functions f and g in Eqs. (7.4) are unknown. But the Poincaré map Σ_P and the impact map Σ_I can be reconstructed from the phase flow (Eqs. 7.1a–b) according to the following relations which define a switch from the previous iteration to the consecutive one of the discussed mappings:

$$\Sigma_P = \begin{cases} x_n = x(t_n) \to x_{n+1} = x(t_n + 2\pi/\omega), \\ u_n = u(t_n) \to x_{n+1} = u(t_n + 2\pi/\omega), \end{cases}$$

(7.5a)

and

$$\Sigma_z = \begin{cases} v_n = v(t_n) \to v_{n+1} = v(t_n + \tau_{n+1}), \\ \theta_n = \cos^{-1}(\omega t) \to \theta_{n+1} = \cos^{-1}[\omega(t + \tau_{n+1})], \end{cases}$$

(7.5b)

where τ_{n+1} is time between consecutive impacts, i.e., iterations n and $n+1$.

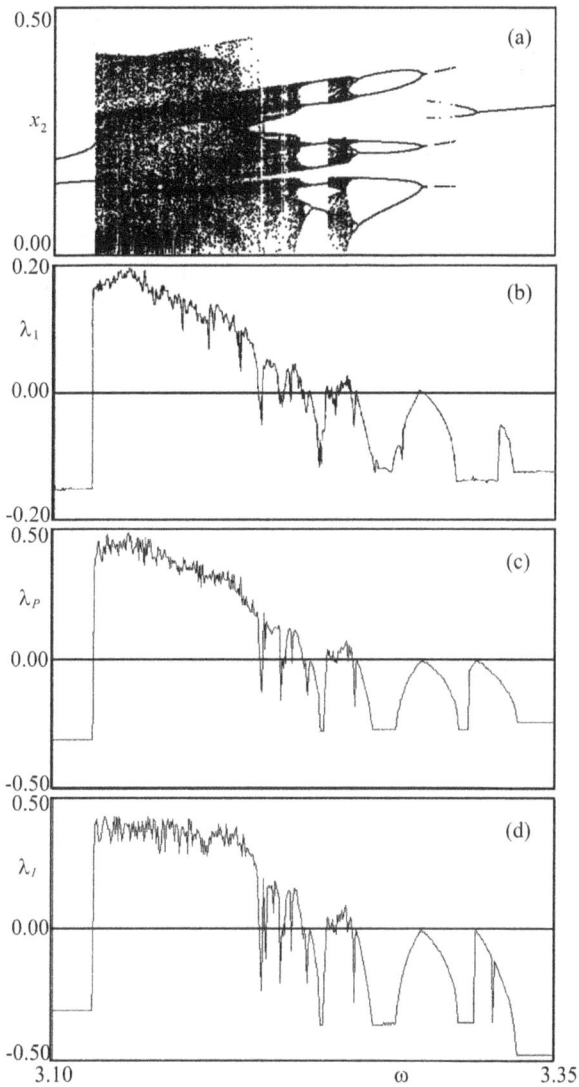

Fig. 7.6. Bifurcational diagram of the mechanical oscillator with impacts (Eqs. (7.1a) and (7.1b)) — velocity x_2 as a function of the excitation frequency ω (a), and the related traces of the LLE estimated with *variant* 1 of the synchronization method (b), and with *variant* 3 for the Poincaré map (c), and the impact map (d); $\alpha=1.00$, $h=0.10$, $p=1.00$, $R=0.90$, $\delta_0=0.00$; numerical procedure parameters: $\varepsilon=0.001$, $D=3.00$, $n_t=500$, $n_e=500$, $d_{1(0)}=-0.80$, $d_{2(0)}=0.80$, $\delta=0.005$.

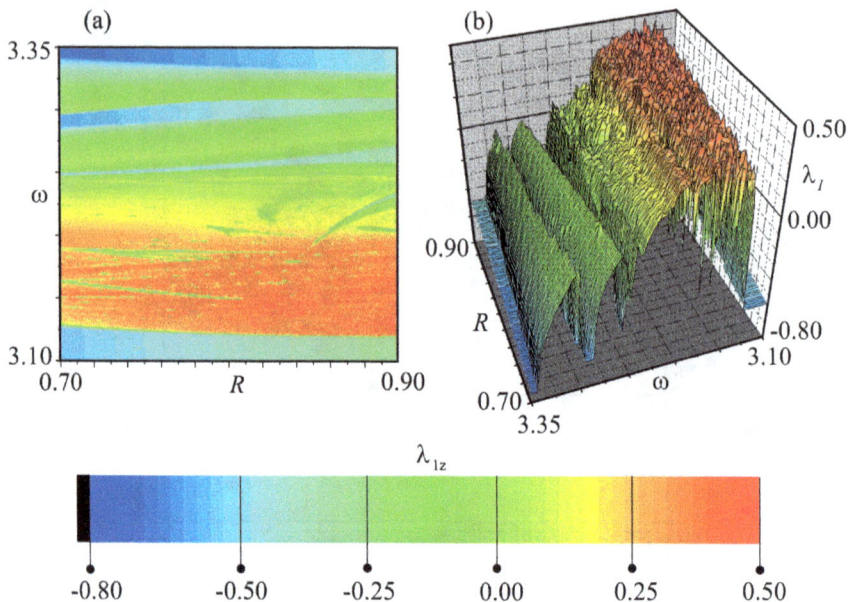

Fig. 7.7. LLE of the mechanical oscillator with impacts (Eqs. (7.1a) and (7.1b)) estimated by *variant* 3 of the synchronization method based on the impact map, shown as a coloured map (a), and a 3-D perspective (b), as a function of the two parameters ω and R; h=0.10, p=1.00, δ_0=0.00; numerical procedure parameters: ε=0.001, D=3.00, n_t=500, n_e=500, $d_{1(0)}$= –0.80, $d_{2(0)}$=0.80, δ=0.005.

Substituting the analyzed maps (Eqs. (7.4)) into Eqs. (6.11a–c), we obtain the augmented system in the following form:

$$v_1^* = f(v_2+\Delta v, \xi_2+\Delta\xi),$$
$$\xi_1^* = g(v_2+\Delta v, \xi_2+\Delta\xi),$$
$$v_2^* = f(v_2,\xi_2),$$
$$\xi_2^* = g(v_2,\xi_2),$$
$$\Delta v^* = [f(v_2+\Delta v,\xi_2+\Delta\xi)-f(v_2,\xi_2)]\exp(-d),$$
$$\Delta\xi^* = [g(v_2+\Delta v,\xi_2+\Delta\xi)-g(v_2,\xi_2)]\exp(-d),$$

(7.6)

which is related to the realization of *variant* 3 of the estimation method. Figures 7.5 and 7.6 present results of the estimation of the LLE with this method in three ways: according to *variant* 1 (Figs. 7.5a and 7.6b) of the method with Eqs. (7.3), according to *variant* 3 of the method for the Poincaré map (Figs. 7.5b and 7.6c), and, the impact map (Figs. 7.5c and 7.6d) with Eqs. (7.4)–(7.6). The estimated value of the λ_1 exponent is presented in top parts of the phase portraits and the maps in Figs. 7.5a–c, and also shown as bifurcation diagrams of the above-mentioned exponent as a function of the excitation frequency Ω (Figs. 7.6b–d), which are related to the bifurcation diagram (Fig. 7.6a) of the system (7.1a–b) velocity x_2. The symbols λ_P and λ_I stand for the LLE of the Poincaré map and the impact map, respectively. Differences in the λ_1 value seen there, sometimes quite large, result from Eq. (5.31), i.e., from the relations between the Lyapunov exponents of the phase flows and the LE of its mappings.

The colored maps presented in Figs. 7.7a and b allow for distinguishing the fractal structure of diffused areas of chaos and regular motion in the two-dimensional parameter space — Ω and R of the impact oscillator (Eqs. 7.1a–b). The maps of the above regimes were constructed on the basis of values of the LLE estimated with *variant* 3 for the impact map (Eqs. 7.4b).

7.2.2 1-DoF stick-slip oscillator

The second example of the non-smooth system is a typical stick-slip oscillator shown in Fig. 7.8. A mass on the conveyor belt, moving with the constant velocity v_B, is kinematically excited via a spring of the stiffness k. The dynamics of the system can be described by the following second order differential equation:

$$m\ddot{x} = -k(x - U_0 \cos \Omega t) + \varepsilon F_N f_\alpha \text{sign}(v), \qquad (7.7)$$

where m — mass of the oscillator, Ω — excitation frequency, U_0 — excitation amplitude, v is the relative velocity (i.e., $v = \dot{x} - v_B$), F_N is the normal pressure force, f_α — friction function, where α denotes an assumed friction model, i.e., the Coulomb model (Eq. (7.8)) or the Popp–Stelter characteristic (Eq. (7.9)), and ε is a non-dimensional constant value responsible for the normal force.

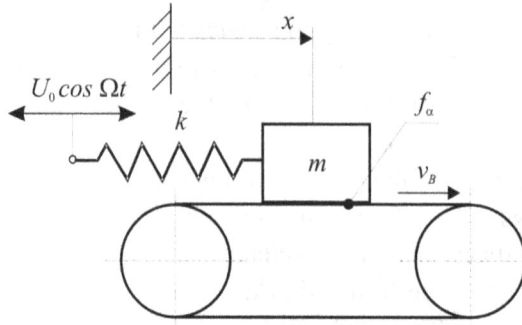

Fig. 7.8. 1-DoF stick-slip oscillator

A practical engineering approach, indebted to Coulomb (Coulomb (1785)), simplifies the friction force during slip to a constant value being opposite to the relative velocity of the contacting bodies. It is expressed as:

$$F_f = F_N f_C \text{sign}(v),$$ (7.8)

where F_f is a friction force, F_N — normal load, f_C — coefficient of Coulomb friction. This is the simplest static model of friction — see Fig. 7.9a. It can distinguish stick and slip phases, assuming that during the stick ($v=0$) the friction coefficient f_S to be different (higher) than during slip.

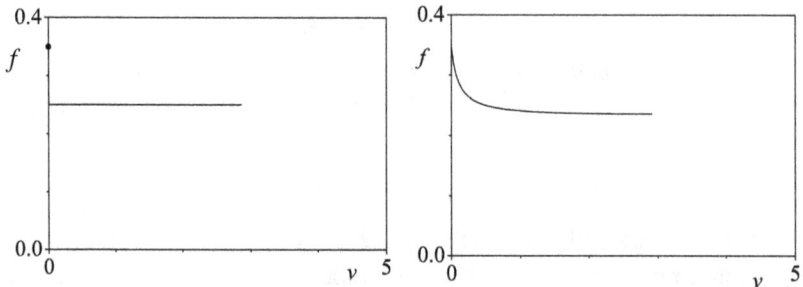

Fig. 7.9. Coulomb (a) and Popp–Stelter (b) friction characteristics generated from system (7.7).

On the other hand, the Popp–Stelter model is characterized by the nonlinear friction–velocity dependence having the constant limit at $v=0$

and the quadratic nonlinearity for $v \neq 0$ — see Fig. 7.9b. Such nonlinear characteristic is represented by the friction coefficient described as follows:

$$f = \frac{f_S - f_C}{1.0 + \eta_1 |v|} + f_C + \eta_2 v^2, \tag{7.9}$$

where η_1, η_2 are constants. This approach reflects the character of the friction–velocity dependence in a more realistic way.

A distinction between the stick and slip phases and back is made by monitoring the restoring force and the maximum static friction. An occurrence of the stick phase is governed by the following inequality:

$$k(x(t) - U_0(t)) < \varepsilon F_N f_S. \tag{7.10}$$

Introducing $\omega = \sqrt{k/m}$ (natural frequency), $x_S = F_N/k$ (static deflection), and then dividing Eq. (7.7) by the factor kx_s, Eq. (7.7) is transformed into a dimensionless form of the first order deferential equations:

$$\begin{aligned}
\dot{x}_1 &= x_2, \\
\dot{x}_2 &= -x_1 + u_0 \cos(x_3) + \varepsilon f_\alpha \mathrm{sign}(v), \\
\dot{x}_3 &= \eta,
\end{aligned} \tag{7.11}$$

where:

$$\tau = \omega t, \quad \eta = \frac{\Omega}{\omega}, \quad u_0 = \frac{U_0}{x_S}, \quad v = (v_B - \dot{x})\frac{1}{\omega x_S},$$

$$x_1 = \frac{x}{x_S}, \quad x_2 = \frac{dx_1}{d\tau} = \frac{dx}{d\tau}\frac{1}{\omega x_S}, \quad x_3 = \eta\tau, \quad \dot{x}_2 = \frac{d^2 x}{dt^2}\frac{1}{\omega^2 x_S}$$

are nondimensional parameters and variables.

During the stick phase, Eqs. (7.11) are reduced to the form:

$$\dot{x}_1 = v_0, \quad \dot{x}_2 = 0, \quad \dot{x}_3 = \eta, \tag{7.12}$$

where $v_0 = v_B / (\omega x_s)$ is the dimensionless belt velocity.

In all numerical computations presented here, we took the following values of parameters: $u_0 = 0.5$, $v_0 = 1.0$, $f_S = 0.35$, $f_C = 0.20$, $f_C = 0.23$, $\eta_1 - 10.0$, and, $\eta_2 = 0.00025$. The frequency η was varied during our numerical studies.

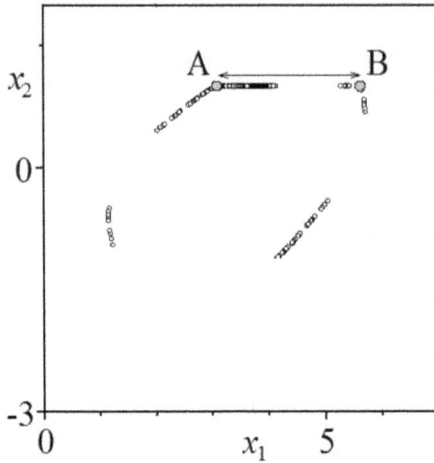

Fig. 7.10. Poincaré map of the stick-slip friction oscillator (Eqs. (7.11 and 7.12)).

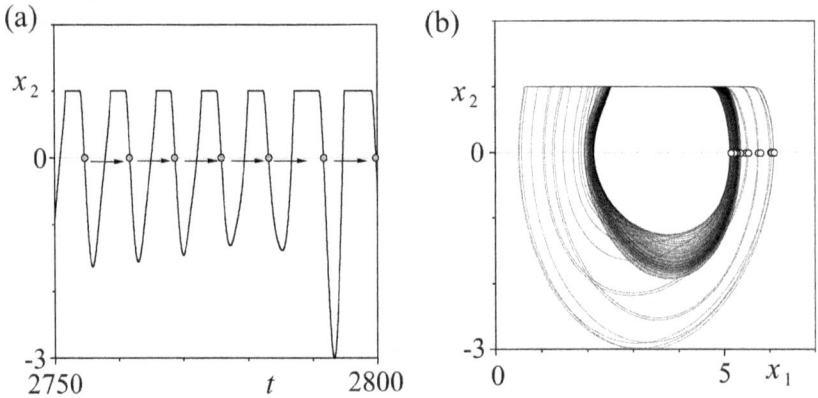

Fig. 7.11. Construction of the map $x_2 = 0$ presented on the velocity time history (left) and on the phase space (right).

In order to implement the synchronization method of the LLE estimation for the considered friction oscillator, its *variant* 3 can be applied. Then, an arbitrary map of the system motion has to be generated. The simplest solution seems to be the Poincaré map obtained by sampling a time history with the period of the harmonic driving $2\pi/\eta$ — see Fig. 7.10. Since the stick phase can be relatively long, consecutive Poincaré points would occur during the transition through the discontinuity (points located on the segment AB of the map in Fig. 7.10), which is characterized by a simple, constant velocity dynamics (Eq. (7.12)). In such a case, an accidental synchronization of the reference and auxiliary maps may appear, which can cause serious estimation errors. The same problem can appear during the possible application of *variant* 1 of the method in the case under analysis. Therefore, we apply the mapping which allows us to avoid the transition through the stick phase.

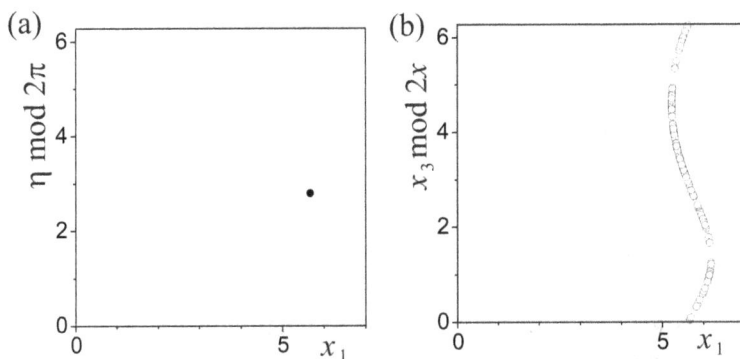

Fig. 7.12. Reconstructed map for $x_2 = 0$; (a) periodic motion for $\eta = 1.72$, $\lambda_1 = -0.199$; (b) chaotic motion for $\eta = 1.79$, $\lambda_1 = +0.016$. The LLE estimated with the synchronization method — top of the drawings; numerical procedure parameters: $\varepsilon=0.001$, $D=1.00$, $n_t=500$, $n_e=500$, $d_{1(0)}=-0.30$, $d_{2(0)}=0.30$, $\delta=0.005$.

The maps were reconstructed for the consecutive moments when the velocity change its sign (from "+" to "–"), i.e., the system trajectory crosses downwards the zero velocity — see Figs. 7.11a–b. As a result, we obtained the 2-D maps shown in Figs. 7.12a–b, which can be described in the following general form:

$$x_1^{n+1} = f\left(x_1^n, x_3^n\right),$$
$$x_3^{n+1} = g\left(x_1^n, x_3^n\right),$$

(7.13a)

where x''_1 and x''_3 represent the displacement of the mass and the phase of harmonic driving at $x_2 = 0$, respectively. Substituting Eq. (7.13a), which plays a role of the reference map, into Eqs. (6.11a) and (6.11c), we obtain the augmented triple system (7.13a–c) with:

an auxiliary disturbed map:

$$y_1^{n+1} = f\left(x_1^n + z_1^n, x_3^n + z_3^n\right),$$
$$y_3^{n+1} = g\left(x_1^n + z_1^n, x_3^n + z_3^n\right),$$

(7.13b)

and
a synchronization error map:

$$z_1^{n+1} = \left(y_1^{n+1} - x_1^{n+1}\right)\exp(-d),$$
$$z_3^{n+1} = \left(y_3^{n+1} - x_3^{n+1}\right)\exp(-d),$$

(7.13c)

in the appropriate form to be used in the estimation procedure realized according to *variant* 3 of the method.

The results of the more detailed study of the analyzed system for both friction models (Eqs. (7.8) and (7.9)) are demonstrated in Figs. 7.13a and 7.13b. Here, the bifurcation diagrams of the variable x_1 and the corresponding LLEs drawn versus the driving frequency η are compared. The LLEs have been evaluated with *variant* 3 of the method using maps (7.13a–c) and, next recalculated according to formula (5.31). Thus, the given LLEs correspond to the phase flow of the friction oscillator. The governing mechanism of changes from periodic to irregular motion looks alike in both bifurcation diagrams. The course of the estimated values of the LLE for both friction models is similar and fully reflects changes in the bifurcation diagrams and presents the existing slight differences between selected models as a result of simulations. Sections of periodic motions on the bifurcation diagrams fully cover the analogous ranges of negative values of the LLE, and *vice versa*, positive ranges of the LLE reflect chaotic ranges in the bifurcation diagrams. Thus, we can state that the synchronization method works equally well in each of the considered models.

(a)

(b)

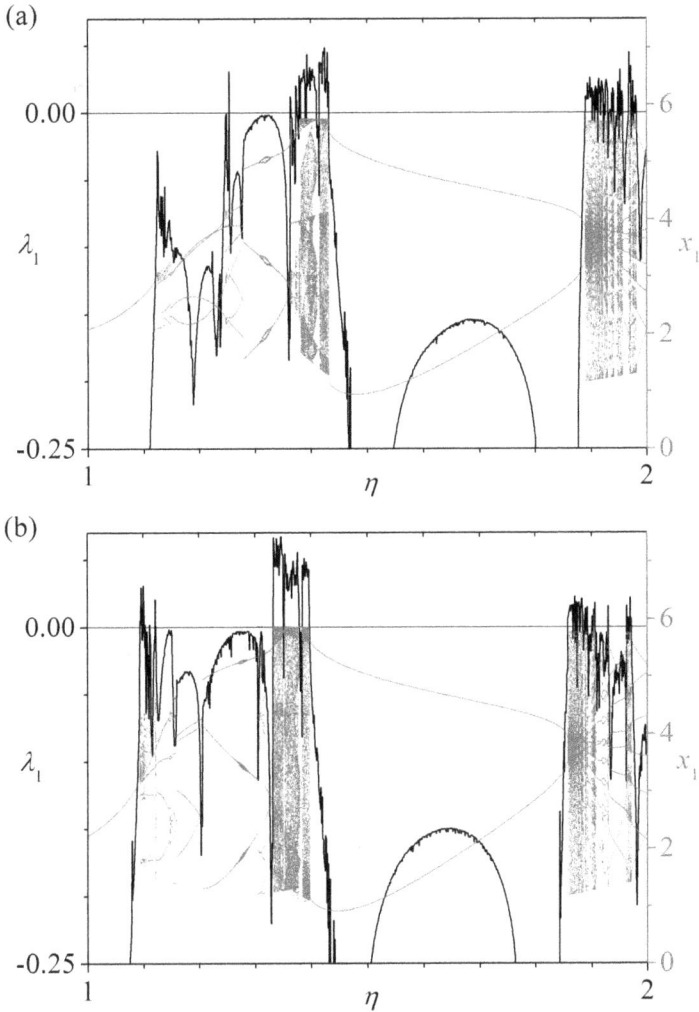

Fig. 7.13. Bifurcation diagrams of system (7.11) – (7.12) (in gray), and the corresponding LLE (black lines) for: (a) Coulomb model, (b) Popp–Stelter friction characteristics; numerical procedure parameters: $\varepsilon=0.001$, $D=1.00$, $n_t=500$, $n_e=500$, $d_{1(0)}=-0.30$, $d_{2(0)}=0.30$, $\delta=0.005$.

7.2.3 2-DoF mechanical oscillator with impacts and friction

Another discussed system with discontinuities (Blazejczyk – Okolewska

(1995), Blazejczyk – Okolewska & Kapitaniak (1996)) is presented in Fig. 7.14. The system consists of the mass m_1 connected to a spring with the stiffness k_1 and a viscous damper characterized by the coefficient c_1. The harmonic excitation described with $P_0\cos(\omega t)$ is applied to the mass m_1. Another mass m_2 is located on it, and its position relative to m_1 is limited with bumpers A and B. The friction force between masses F_T and the viscous damping force with the coefficient c_2 are also taken into account. The mechanical model with 2-DoF in time between impacts can be described by the following first order differential equations:

$$\dot{y}_1 = y_2,$$
$$\dot{y}_2 = p\cos(\eta t) - b_1 y_2 - y_1 + f_T - b_1\sigma(y_2 - y_4),$$
$$\dot{y}_3 = y_4,$$
$$\dot{y}_4 = (f_T/\gamma) + (\sigma/\gamma)(y_2 - y_4),$$

(7.14)

where: $\tau = \alpha_1 t$, $\alpha_1 = \sqrt{k_1/m_1}$, $\eta = \omega/\alpha_1$, $x_{st} = m_2 g/k_1$, $b_1 = c_1/\alpha_1$, $y_1 = x_1/x_{st}$, $y_2 = x_1$, $y_3 = x_2/x_{st}$, $y_4 = x_2$, $\gamma = m_2/m_1$, $\sigma = c_2/c_1$, $f_T = F_T/m_2 g$, $p = P_0/m_2 g$ (variables and parameters as in Fig. 7.14).

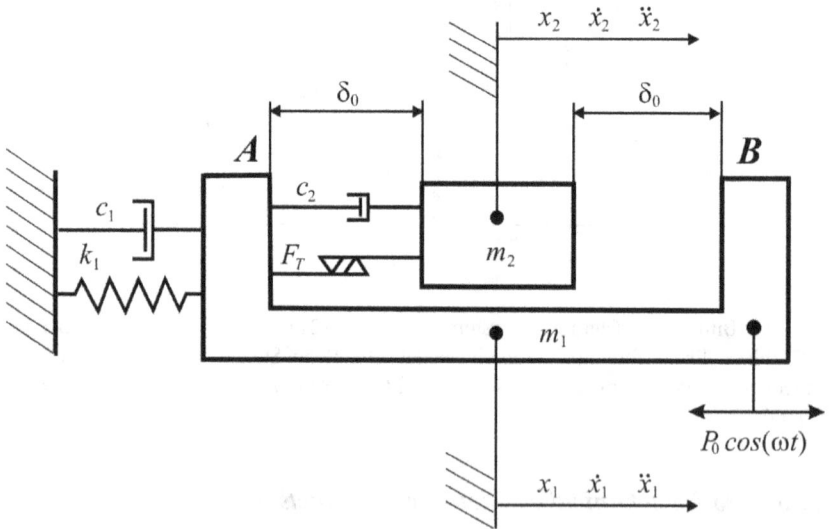

Fig. 7.14. Model of the 2-DoF mechanical oscillator with motion limits (bumpers A and B).

Impact conditions of the masses are described by the following relations:

$$|y_3 - y_1| \geq l,$$ (7.15)

where: $l = \delta_0 / x_{st}$, and,

$$y_2^+ = \frac{m_1 y_2^- + m_2 y_4^-}{m_1 + m_2} - \frac{m_2}{m_1 + m_2} R(y_2^- - y_4^-),$$ (7.16a)

$$y_4^+ = \frac{m_1 y_2^- + m_2 y_4^-}{m_1 + m_2} + \frac{m_1}{m_1 + m_2} R(y_2^- - y_4^-),$$ (7.16b)

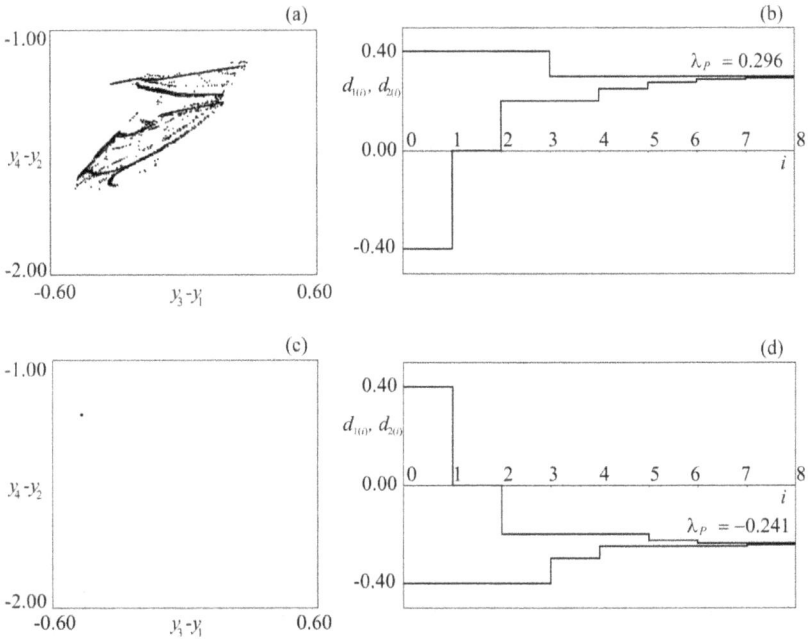

Fig. 7.15. Poincaré maps ((a) and (c)) of the 2-DoF mechanical oscillator with motion limits (Eqs. (7.14) (7.17)) and the related traces of the LLE estimation results obtained by the bisection method during 8 iterations ((b) and (d)); chaotic motion ((a) and (b)) for η=1.23, $1T$ periodic motion ((c) and (d)) for η=1.44. System parameter values: γ=0.693, σ=0.50, μ=0.02, b_1=0.10, l=0.80, p=1.00, R=0.60. Numerical procedure parameters: ε=0.005, D=2.00, n_t=1500, n_e=1500, $d_{1(0)}$=–0.40, $d_{2(0)}$=0.40, δd=0.01.

where the signs "–" and "+" are related to the velocities before and after the impact, respectively, and R is the restitution coefficient. In Eqs. (7.14), the linear Coulomb friction was assumed. It is described by:

$$f_T = \mu \, \mathrm{sgn}(y_2 - y_4), \qquad (7.17)$$

where μ is the friction coefficient of the contact surface.

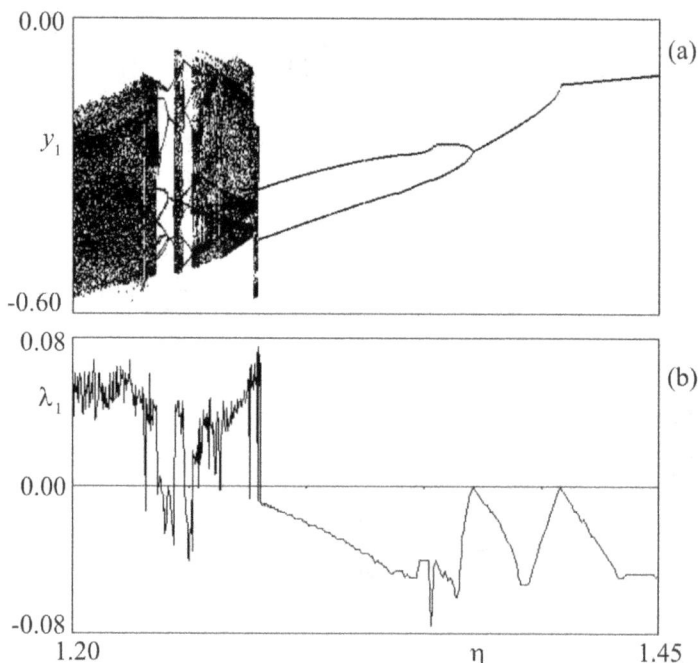

Fig. 7.16. Bifurcation diagram of the 2-DoF mechanical oscillator with motion limits (Eqs. (7.14) – (7.17)), as a function of the dimensionless driving frequency η (a), and the corresponding trace of the LLE estimated according to *variant* 3, based on the Poincaré map (b); γ=0.693, σ=0.50, μ=0.02, b_1=0.10, l=0.80, p=1.00, R=0.60; numerical procedure parameters: ε=0.01, D=2.00, n_r=500, n_e=500, $d_{1(0)}$= -0.40, $d_{2(0)}$=0.40, δd=0.01.

The estimation of the LLE of system (7.14 – 7.17) was performed with *variant* 3 of the estimation method based on the Poincaré map. In this case any use of *variant* 1 may lead to considerable errors in the results, due to the fact that both masses may stick in some phase of the motion, which in presence of a coupling, can result in immediate synchronization of the coupled, twin type phase flows, and, consequently, in a poor final result of the estimation process. On the other hand, in the case of Poincaré maps generated from differential equations, a probability of their occasional synchronization, even while they stick, is very small. Such a map for system (7.14 – 7.17) is a 4-D cross-section of its 5-D phase space.

According to Eqs. (6.11a–c), the realization of *variant* 3 of the method in the system under consideration requires the determination of twelve variable values $x_i = y_i + z_i$, y_i, z_i (i=1, 2, 3, 4) in the process of the LLEs estimation from the scheme shown in Fig. 6.6. Poincaré maps and the related "history" of the LLE estimation with the bisection method are shown in Figs. 7.15a–d. Exemplary estimated values of the exponent λ_P, characterizing the Poincaré maps from Figs. 7.15a and 7.15c, are given in the neighboring Figs. 7.15b and 7.15d, in the case of the chaotic and periodic solution, respectively. The bifurcation diagram for system (7.14–7.17) and the corresponding course of the LLE versus the frequency η are presented in Figs. 7.16a and 7.16b, respectively. Here, the LLE λ_1 has been recalculated according to Eq. (5.31), i.e.,

$$\lambda_1 = \frac{\eta \lambda_P}{2\pi}, \tag{7.18}$$

because the period of harmonic driving $T = 2\pi/\eta$ is a constant time between two consecutive iterations of the Poincaré map.

7.2.4 3-DoF impact absorber of vibration

The next analyzed example is a mechanical system with discontinuities which can work as a vibration damper with motion limiters (Dabrowski (2002)). This is a 3-DoF system with an external harmonic driving (see Fig. 7.17). A source of discontinuities in the system are impacts between the masses m_1 and m_2. Its mathematical description is a set of six first order differential equations:

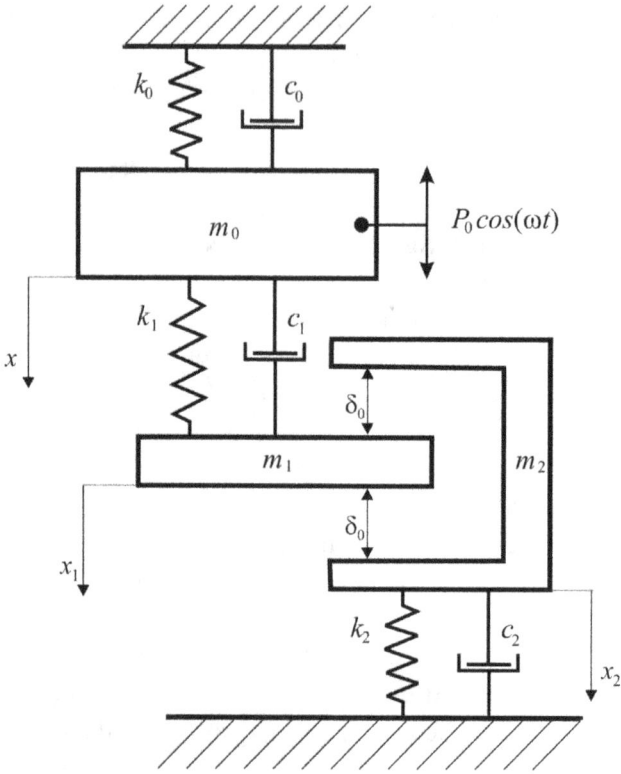

Fig. 7.17. Model of the vibration damper with motion limiters.

$$\dot{y}_1 = y_2,$$
$$\dot{y}_2 = p\sin(\eta t) - 2\gamma_0 y_2 - y_1 + 2\gamma_1 \sigma_1 \beta_1 (y_4 - y_2) + \sigma_1 (y_3 - y_1),$$
$$\dot{y}_3 = y_4,$$
$$\dot{y}_4 = -2\gamma_0 \mu_1^{-1} y_2 + 2\gamma_1 \beta_1^{-1} (y_2 - y_4) + \beta_1^{-2} (y_1 - y_3), \tag{7.19}$$
$$\dot{y}_5 = y_6,$$
$$\dot{y}_6 = -2\gamma_2 \beta_2^{-1} y_6 - \mu_2^{-1} y_5 / \mu_2 - \beta_2^{-2} (y_5 + \delta_0),$$

while conditions of impacts of the masses m_1 and m_2 are defined analogously as in the previous 2DoF example (Eqs. (7.15) and (7.16a–b)).

Equation (7.19) were obtained after the following substitutions: $y_1 = x$, $y_2 = \dot{x}$, $y_3 = x_1$, $y_4 = \dot{x}_1$, $y_5 = x_2$, $y_6 = \dot{x}_2$ and $\tau = \alpha_0 t$, $\alpha_0 = \sqrt{k_0/m_0}$, $\eta = \omega/\alpha_0$, $p = P_0/k_0$, $\mu_1 = m_1/m_0$, $\mu_2 = m_2/m_0$, $\sigma_1 = k_1/k_0$, $\sigma_2 = k_2/k_0$, $\gamma_0 = c_0/\left(2\sqrt{k_0/m_0}\right)$, $\gamma_1 = c_1/\left(2\sqrt{k_1/m_1}\right)$, $\gamma_1 = c_1/\left(2\sqrt{k_1/m_1}\right)$, $\beta_1 = \sqrt{\mu_1/\sigma_1}$, $\beta_2 = \sqrt{\mu_2/\sigma_2}$ (variables and parameters as in Fig. 7.17).

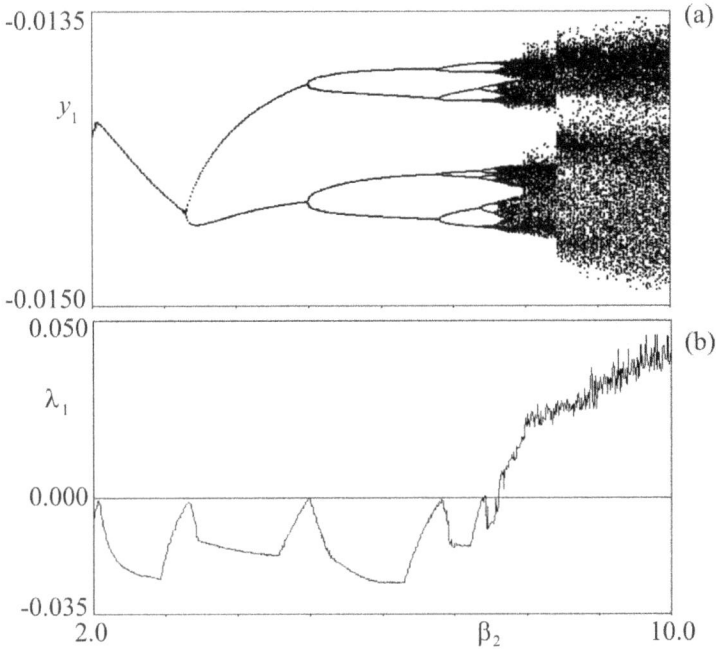

Fig. 7.18. Bifurcation diagrams for the vibration damper with motion amplitude limiters (Eqs. (7.19)), as a function of the parameter β_2 (a), and the corresponding LLE estimated according to *variant* 3 on the basis of Poincaré maps (b); β_1=1.25, γ_0=γ_1=0.02, γ_2=0.10, σ_1=0.10, σ_2=1.50, μ_1=μ_2=0.10, δ_0=0.015, η=1.07, p=0.002, R=0.50; numerical procedure parameters: ε=0.001, D=1.00, n_t=300, n_e=500, $d_{1(0)}$=–0.40, $d_{2(0)}$=0.40, δd=0.005.

In the case of the analyzed system, the estimation of the LLE has been also performed with *variant* 3 of the method on the basis of the Poincaré map. This map is defined by 6 variables in system (7.19), which are reconstructed every period of the excitation force, analogously to the non-smooth impact–friction oscillator (Eqs. (7.14)). Hence, values of the LLE in all the cases presented here have been computed with use of Eq. (7.18).

The results of the estimation are shown in the form of a comparison of bifurcation diagrams and the related courses of the exponent λ_1. They are depicted in Figs. 7.18, 7.19 and illustrated in colors (Fig. 7.20) to show values of the LLE in the 2-D parameter space of system (7.19).

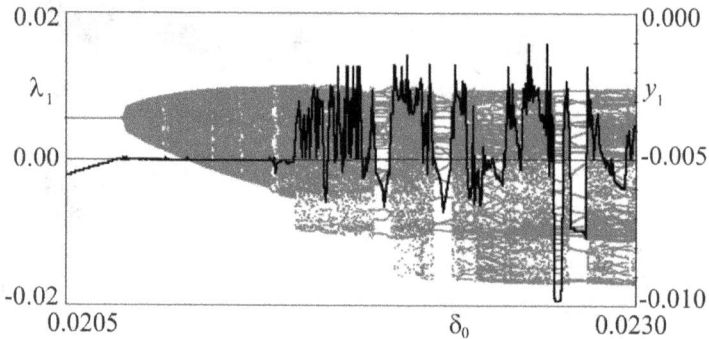

Fig. 7.19. Bifurcation diagram of the vibration damper with motion amplitude limiters (Eqs. (7.19)), as a function of the parameter δ_0 (in gray), and the corresponding LLE estimated according to *variant* 3 of the method using the Poincaré map (in black); β_1=1.00, β_2=1.00, γ_6=0.02, γ_1=0.05, γ_2=0.10, σ_1=0.10, σ_2=1.50, μ_1=μ_2=0.10, η=1.165, p=0.002, R=0.50; numerical procedure parameters: ε=0.0002, D=1.00, n_i=500, n_e=500, $d_{1(0)}$=-0.20, $d_{2(0)}$=0.20, δd=0.005.

To summarize Sec. 7.2, we can state that the presented above examples of the estimation of the LLE for non-smooth mechanical systems using the synchronization method prove its efficiency regardless the number of degrees of freedom of the investigated system and the type of discontinuity (impact, friction). A quality of the results is also proven by a comparison of the bifurcation diagrams and the corresponding traces of the λ_1 exponent presented in Figs. 7.6, 7.7, 7.13, 7.16 and 7.18 – 7.20. Full agreement of the estimation results compared to the observed type of the dynamical behavior, i.e., the exponent λ_1 has a positive value in chaotic regimes; zero in the case of quasi-periodic motion (see Fig. 7.19) and a negative one in periodic regimes, can be easily seen there. The comparison also shows that the estimation results allow for an identification of basic types of local bifurcations of the local periodic solutions, as the period doubling, the Hopf or saddle–node bifurcation.

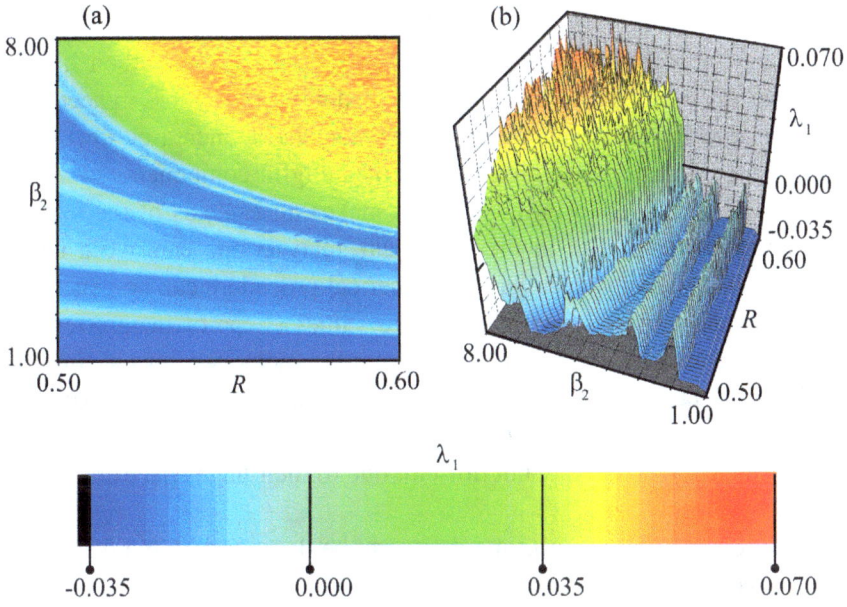

Fig. 7.20. LLE for the vibration damper with motion amplitude limiters (Eqs. (7.19)) estimated with *variant* 3 of the method, presented in the form of a colored map (a), and a 3-D perspective view (b), as a function of the two parameters R and β_2; $\beta_1=1.25$, $\gamma_0=\gamma_1=0.02$, $\gamma_2=0.10$, $\sigma_1=0.10$, $\sigma_2=1.50$, $\mu_1=\mu_2=0.10$, $\delta_0=0.015$, $\eta=1.07$, $p=0.002$; numerical procedure parameters: $\varepsilon=0.001$, $D=1.00$, $n_f=300$, $n_e=500$, $d_{1(0)}=-0.40$, $d_{2(0)}=0.40$, $\delta d=0.005$.

7.3 Systems with Time Delay

Phase flows with time delay can be in general described with the following matrix equation:

$$\dot{\mathbf{x}} = \mathbf{f}(\mathbf{x}(t), \mathbf{x}(t - \tau_0)),$$ (7.20)

where $\tau_0 \in \Re$ represents the time delay. Generally, a presence of the time delayed variable $\mathbf{x}(t-\tau_0)$ does not exclude the use of algorithms for calculating Lyapunov exponents based on the Oseledec theorem. The expression $\mathbf{x}(t-\tau_0)$ can be transformed into a Taylor series in the neighborhood of the τ_0 value according to the formula:

$$\mathbf{x}(t - \tau_0) = \mathbf{x}(t) + \frac{1}{i!}\sum_i(-\tau_0)^i\frac{d^i\mathbf{x}(t)}{dt^i}. \qquad (7.21)$$

When the delay parameter τ_0 remains relatively small, then higher order terms in series (7.21) can be neglected without any loss in accuracy of the solution. However, in the case when the τ_0 increases up to the value of 1 or more, the number of the meaningful components of series (7.21) also increases to infinity, which makes an analysis of the system extremely difficult. But in the numerical analysis one can avoid using formula (7.21), due to fact that the numerical procedure can be generated to "remember" the consecutive discrete values (steps of integration) of the delayed variable, and recall them from the memory after the time τ_0, during the process of the numerical integration of Eq. (7.20). An application of the classical methods for calculating the Lyapunov exponents is possible here, but difficult due to their huge spectrum in this case.

Discrete mappings with time delay can be described by the following difference equation:

$$\mathbf{x}_{n+1} = \mathbf{f}\left(\mathbf{x}_n,\mathbf{x}_{n-n_0}\right) \qquad (7.22)$$

where $n_0 \in N$ labels a map delay. In this case, a numerical reconstruction of the consecutive iteration of Eq. (7.22) can be performed by recalling the stored, delayed values of $\mathbf{x}_{n-n0}$, which, in practice, results in the analysis of system (7.22) in the form extended with an n_0 number of equations defining evolution of the delayed variable, which can be written as a set of the following difference equations:

$$\left\{\mathbf{x}_{n-j} = \mathbf{x}_{n-j+1}\right\} \qquad (7.23)$$

where: $j=1, 2, 3, ..., n_0$.

In the dynamical systems without any delay, the disturbance caused by a coupling yields already a result in the next iteration. When delay is considered, its effect does not appear in the next iteration, but after the time defined by the parameters τ_0 or n_0. However, such a difference does not change properties of the synchronization mechanism of identical systems (6.1a–b) or (6.11a–b) due to the uniform CD coupling introduced between them. The condition of synchronization defined with inequality (6.8) is fulfilled also in the dynamical systems with delay, which allows for estimating the largest Lyapunov exponent of such systems with the synchronization method. Practically, during the estimation process of the values of λ_1 in systems with delay, the method is realized in the same manner as in other cases, i.e., by substituting (7.20) or (7.22) into (6.1a–b) or (6.11a–c), respectively.

Now let us consider the estimation of the LLE for the following dynamical systems with time delay:

1. Henon map as an example of the discrete mapping,
2. Van der Pol oscillator as an example of the phase flow.

7.3.1 *Henon map with time delay*

After substituting the difference equations of the Henon mapping to general form (7.22), the system takes a form of the discrete map with time delay described by:

$$x_{n+1} = 1 - ax_n^2 + y_{n-n_0},$$
$$y_{n+1} = bx_n,$$

(7.24)

where the variable y plays a role of the delayed variable n_0. The estimation of the largest Lyapunov exponent of system (7.24) has been performed with use of expanded system (6.11a–c), analogously to the case of the classical Henon mapping without delay, and the results are shown in Sec. 7.1. In this case, *variant* 3 of the estimation method was applied. Its results are shown in a form of the bifurcation diagrams, which are overlapped with shapes of the λ_1 values for different magnitudes of the delay n_0, see Figs. 7.21a–c.

7.3.2 Van der Pol oscillator with time delay

A nonlinear Van der Pol oscillator with harmonic excitation and time delay (Awrejcewicz & Wojewoda) is described by the following differential equation:

$$\ddot{y} - \alpha(1 - y^2)\dot{y} + \beta y^3 = \kappa_0 y(t - \tau_0) - p\cos(\eta t). \qquad (7.25)$$

In Eq. (7.25), time delay is represented by the first component on the RHS, and its value and level of influence on the system are described with the constant values τ_0 and κ_0.

In case of system (7.25), the synchronization method of estimation of the largest Lyapunov exponent can be realized in *variants* 1 and 3. Simplicity and smoothness of the model with 1DoF makes use of *variant* 1 easy. After changing variables and substituting Eq. (7.25) into Eqs. (6.1a–b), we obtain a double set of four first order, differential equations, which allow for a numerical integration. In this system, *variant* 3 of the method can be realized by the use of the Poincaré maps with care as the integration step Δt should be carefully chosen to have the time delay τ_0 and the period of the excitation force $T=2\pi/\eta$ to be its natural multiplicity. But in the case of the nonautonomous system with delay, as system (7.25), *variant* 3 of the method can be realized by the use of the so-called τ-maps, in which the consecutive iterations are mappings of the phase flow every period of the delay τ_0, and not every period T as in the case of the Poincaré maps. The use of the τ-maps allows for using an arbitrary integration step which is a part of the time delay without adaptation to period T. In the case of system (7.25), the τ-map can be described in a general form:

$$\begin{aligned} y_{n+1} &= f\big(y(t)_n, y(t - \tau_0)_n, \dot{y}(t)_n\big), \\ \dot{y}_{n+1} &= g\big(y(t)_n, y(t - \tau_0)_n, \dot{y}(t)_n\big). \end{aligned} \qquad (7.26)$$

Even if the functions f and g in Eqs. (7.26) are not known, the τ-map can be reconstructed numerically from the phase flow (Eq. (7.25)), according to the relations analogous to Eqs. (7.5a–b):

$$\Sigma_\tau = \begin{cases} y_n = y(t_n) \rightarrow y_{n+1} = y(t_n + \tau_0), \\ \dot{y}_n = \dot{y}(t_n) \rightarrow \dot{y}_{n+1} = \dot{y}(t_n + \tau_0). \end{cases} \qquad (7.27)$$

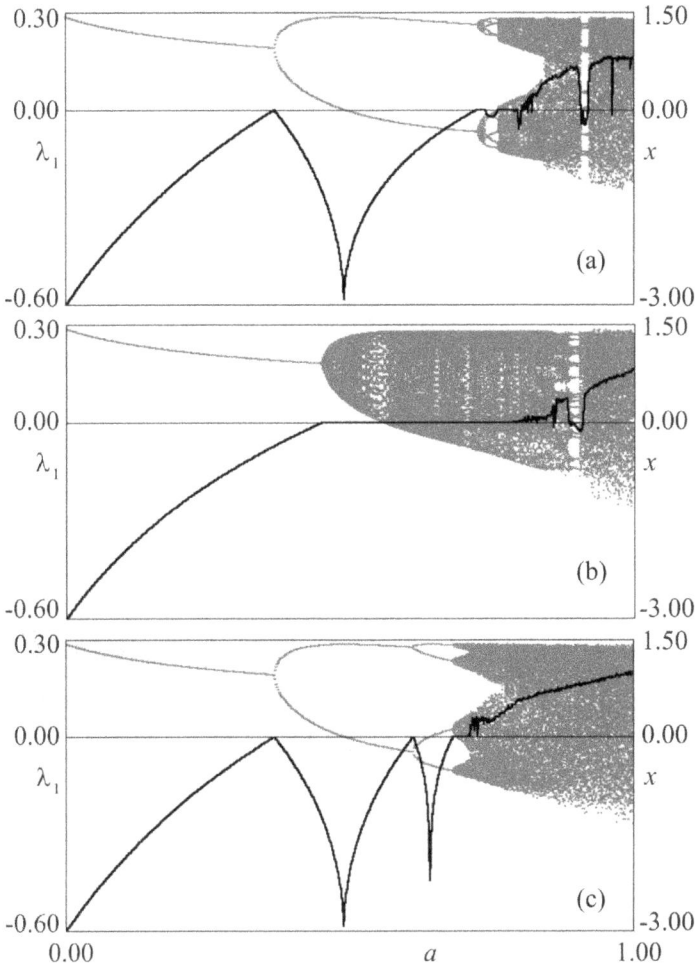

Fig. 7.21. Bifurcational diagrams of the Henon maps with delay (Eq. (7.24)), versus the parameter a (gray), and, the corresponding LLEs estimated with *variant* 3 of the synchronization method (black), for different magnitudes of delay: $n_0=4$ (a), $n_0=6$ (b), $n_0=10$ (c); $b=0.30$. Numerical procedure parameters: $\varepsilon=0.001$, $D=1.00$, $n_t=5000$, $n_e=5000$, $d_{1(0)}=-1.00$, $d_{2(0)}=1.00$, $\delta d=0.001$.

Assuming the time between the consecutive iterations of the τ-map is equal to τ_0 causes that the analysis of the mapping (7.26), reconstructed from the phase flow (7.25), is an analogy of the analysis of the discrete map (7.22) with time delay $n_0=1$, i.e., $y(t-\tau_0)_n = y_{n-1}$. Using this formulation and substituting the analyzed map (Eq. (7.26)) into Eqs. (6.11a–c), we obtain the $3k$-dimensional ($k=2$) system in the following form:

$$x_{n+1} = f\left(y_n + \Delta y, y_{n-1} + \Delta y_{n-1}, \dot{y}_n + \Delta \dot{y}\right),$$
$$\dot{x}_{n+1} = g\left(y_n + \Delta y, y_{n-1} + \Delta y_{n-1}, \dot{y}_n + \Delta \dot{y}\right).$$
$$y_{n+1} = f\left(y_n, y_{n-1}, \dot{y}_n\right),$$
$$\dot{y}_{n+1} = g\left(y_n, y_{n-1}, \dot{y}_n\right), \qquad\qquad (7.28)$$
$$\Delta y_{n+1} = \left[f\left(y_n + \Delta y, y_{n-1} + \Delta y_{n-1}, \dot{y}_n + \Delta \dot{y}\right) - f\left(y_n, y_{n-1}, \dot{y}_n\right)\right]\exp(-d),$$
$$\Delta \dot{y}_{n+1} = \left[g\left(y_n + \Delta y, y_{n-1} + \Delta y_{n-1}, \dot{y}_n + \Delta \dot{y}\right) - g\left(y_n, y_{n-1}, \dot{y}_n\right)\right]\exp(-d),$$

which is related to the numerical realization of *variant* 3 of the estimation method.

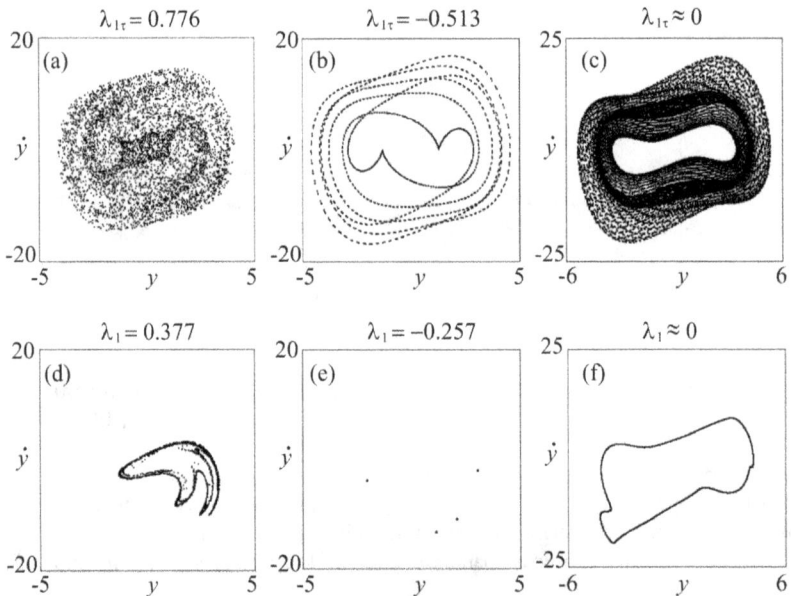

Fig. 7.22. The τ-maps (a), (b), (c) and the corresponding Poincaré maps (d), (e), (f) of the van der Pol oscillator with time delay (Eq. (7.25)) with their LLE (top part) estimated with the synchronization method, *variant* 3 ((a), (b), (c)) and *variant* 1 ((d), (e), (f)); chaotic motion — $\kappa_0=0.80$ (a) and (d), periodic motion — $\kappa_0=2.20$ (b) and (e), quasi-periodic motion — $\kappa_0=6.00$ (c) and (f); $\alpha=0.20$, $\beta=1.00$, $\tau_0=2.00$, $\eta=4.00$, $p=17.00$. Numerical procedure parameters: $\varepsilon=0.1$, $D=4.00$, $n_f=5000$, $n_e=5000$, $d_{1(0)}=-0.50$, $d_{2(0)}=1.00$, $\delta d=0.005$.

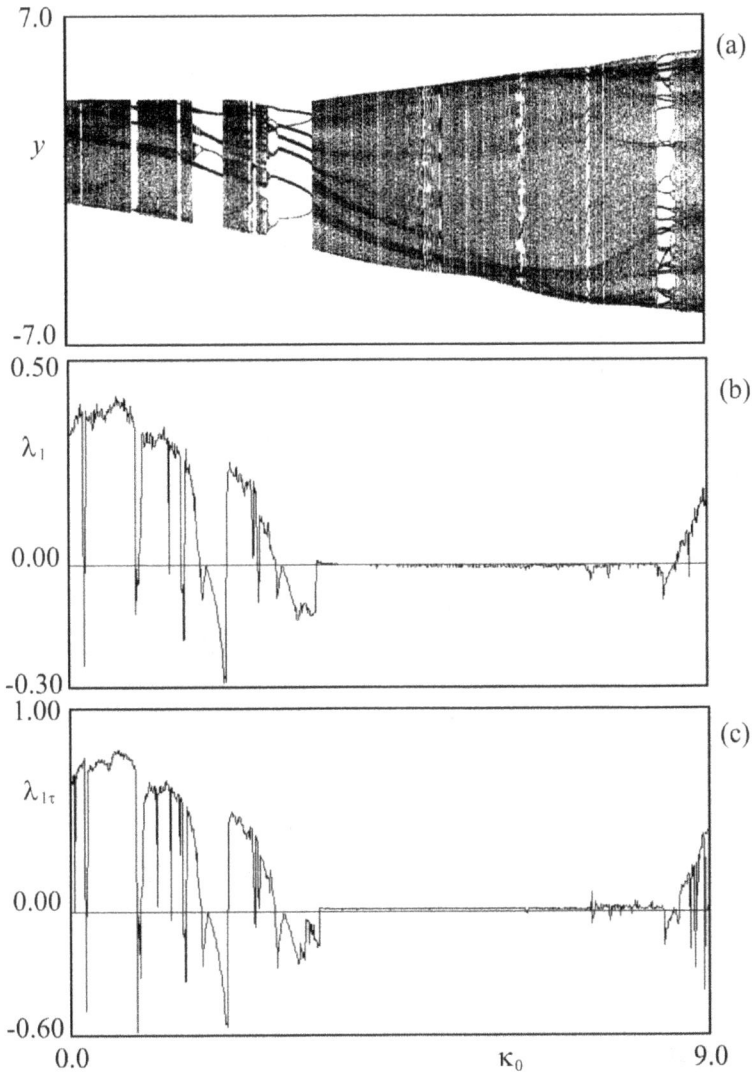

Fig. 7.23. Bifurcation diagram of the variable y of the Van der Pol oscillator with time delay (Eq. (7.25)) versus the parameter κ_0 (a), and the corresponding LLE estimated by *variant* 1 of the synchronization method (b) and *variant* 3 (c); $\alpha=0.20$, $\beta=1.00$, $\tau_0=2.00$, $\eta=4.00$, $p=17.00$. Numerical procedure parameters: $\varepsilon=0.1$, $D=4.00$, $n_f=5000$, $n_e=5000$, $d_{1(0)}=-0.50$, $d_{2(0)}=1.00$, $\delta d=0.005$.

Examples of the τ-maps (Eq. (7.26)) of system (7.25) and their values of the LLE $\lambda_{1\tau}$, which are obtained for these maps with *variant* 3 of the estimation method, are shown in Figs. 7.22a–c. An irrational ratio of the period T to the delay τ_0 makes the τ-maps similar to the phase portraits, while the number of iterations increases. Figures 7.22d–f present the corresponding Poincaré maps and the exponents λ_1 estimated with *variant* 1 of the method. In Fig. 7.23 a bifurcational comparison of the LLEs obtained with *variants* 1 and 3 (Figs. 7.23b and 7.23c, respectively) with the bifurcation diagram of the variable y (Fig. 7.23a) versus the parameter κ_0 is demonstrated. As in the previously discussed examples, the drawings show agreement in the estimation results of the λ_1 values with the observed types of dynamical behavior. A detailed comparison of both *variants* is presented in Sec. 7.5.

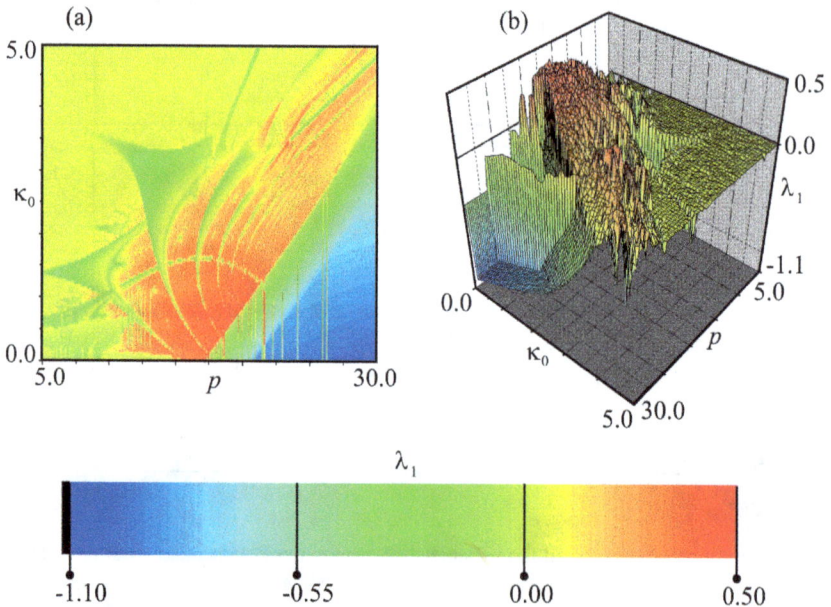

Fig. 7.24. LLE of the Van der Pol oscillator with delay (Eq. (7.25)), estimated with *variant* 1 of the synchronization method, shown as a form of the colored map (a), and a 3-dimensional perspective (b), as a function of the parameters κ_0 and p; α=0.20, β=1.00, τ_0=2.00, η=4.0. Numerical procedure parameters: ε=0.1, D=4.00, n_f=1000, n_e=1000, $d_{1(0)}$= −1.20, $d_{2(0)}$=0.60, δd=0.01.

In Fig. 7.24, a colored presentation of the results of estimation of the λ_1 is depicted, which clearly shows a fractal structure of the surface of the LLE as a function of two parameters of system (7.25). The results were obtained with *variant* 1 of the estimation method, due to the fact that in this case it is much more efficient. But, if some discontinuities appear in the case of the system with delay, *variant* 3 of the method is suggested. In the case of earlier presented examples of estimation of the exponent λ_1 of system (7.25), the delay value is equal to $\tau_0=2.00$.

7.4 Effective Lyapunov Exponents

In this Sec. some examples of determination of the ELE and the ERLE by means of the proposed method are presented.

7.4.1 *ELE estimation*

The numerical experiments have been carried with the use of a mechanical oscillator of the Duffing type with the nonlinear spring characteristics kx^2, the linear viscous friction c and the dynamic harmonic drive of the amplitude F_0 and the frequency Ω (see Fig. 7.25), in which an influence of noise is also considered. The dynamics of this oscillator is governed by the following non-autonomous, dimensionless equations of motion:

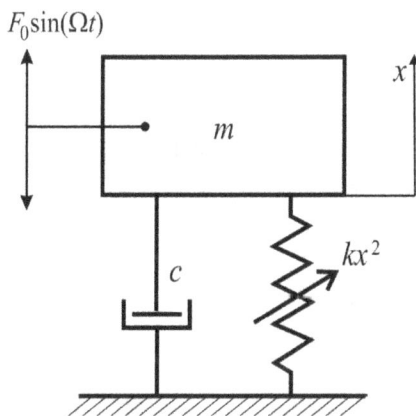

Fig.7.25. Non-linear mechanical oscillator of the Duffing type.

$$\dot{x}_1 = x_2,$$
$$\dot{x}_2 = -\alpha x_1^3 - hx_2 + q\sin(\eta\tau), \tag{7.29}$$

where the parameters α, h, q, η and τ are dimensionless representations of the real parameters k, c, F_0, Ω and the time t, respectively. In all the numerical experiments presented here, $\alpha = 10.0$, $h = 0.3$ and $q = 10.0$ have been taken. In the bifurcation analysis, the frequency of drive η has been used as a control parameter.

In order to estimate the ELE, an auxiliary double system has been constructed:

$$\dot{x}_1 = x_2,$$
$$\dot{x}_2 = -\alpha x_1^3 - hx_2 + q\sin(\eta\tau),$$
$$\dot{y}_1 = y_2 + d(x_1 - y_1), \tag{7.30}$$
$$\dot{y}_2 = -[\alpha + \Delta\alpha(t)]y_1^3 - hy_2 + q\sin(\eta\tau) + d(x_2 - y_2),$$

where $\Delta\alpha$ represents a time-varying mismatch of the parameter α. Hence, the related *mismatch vector* is $\Delta(\mathbf{y}) = [0, \Delta\alpha(t)y_1^3]^{\mathrm{T}}$. As we can see, in the example under consideration, the influence of noise is manifested by the randomly fluctuating parameter α. Such a fluctuation can be generated with numerical techniques modeling any stochastic process, e.g., the Gaussian process by the spectral representation method (Shinozuka (1992)). In the simulations carried out, a random number generator embedded in the DELPHI environment was applied. Time variations of a mismatch parameter are determined with the formula:

$$\Delta\alpha(t) = \rho\, rand[-1, 1] \tag{7.31}$$

where *rand* [-1, 1] is a stochastic function returning a random number uniformly distributed over the interval [-1, 1] in each step of numerical computations. These computations have been carried out employing the RK4 method with a fixed time step $dt = 0.01$ and an amplitude of noise $\rho = 0.03$. The assumed form and the amplitude of noise (Eq. (7.31)) mean that fluctuations of the parameter α are restricted to the range $\pm3\%$ of its magnitude.

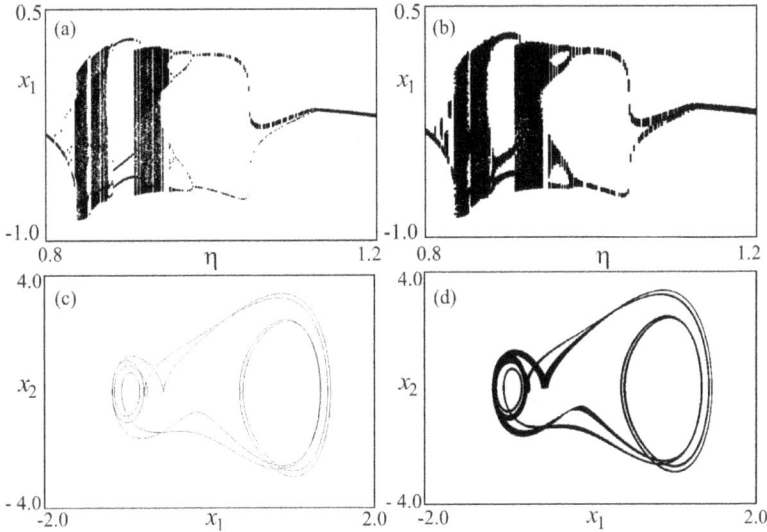

Fig. 7.26. Bifurcation diagrams of the Duffing oscillator (Eq. (7.29)) versus the frequency of the drive η — (a) undisturbed — $\rho = 0$, (b) disturbed with Eq. (7.31) — $\rho = 0.03$, and exemplary phase portraits for $\eta = 0.95$ — (c) without noise, (d) with noise.

The results of the typical bifurcation analysis (the variable x_1 versus the control parameter η) of the considered Duffing oscillator are presented in Figs. 7.26a–d. We can see (Fig. 7.26a) that for smaller values of η, a chaotic motion appears as a result of period-doubling bifurcations ($\eta \approx (0.83, 0.94)$). In the middle of this region, a "periodic window" is observed ($\eta \approx (0.87, 0.91)$). Next, with an increase in the frequency η, chaos disappears in a few sequences of the inverted period-doubling bifurcations. Comparing the bifurcation diagrams depicted in Figs. 7.26a and 7.26b, i.e., computed for the unperturbed ($\rho = 0$) and the perturbed ($\rho = 0.03$) versions of oscillator (7.29), respectively, we can evaluate an influence of the fluctuating parameter α on its dynamics. This influence is clearly visible especially in the intervals of the regular motion (see Fig. 7.26b), where it is manifested by significantly thicker branches of the plot in a comparison with the undisturbed case (Fig. 7.26a). However, the disturbance does not introduce qualitative changes to the global view of the system dynamics (Fig. 7.26b), because an overall structure of the bifurcation plot is kept in spite of noise. The same effect can be observed in the accompanying phase portraits illustrating the dynamics of both cases of the system under consideration (Figs. 7.26c and 7.26d). Here the noise-induced perturbation of the system trajectory is also evident (Fig. 7.26d).

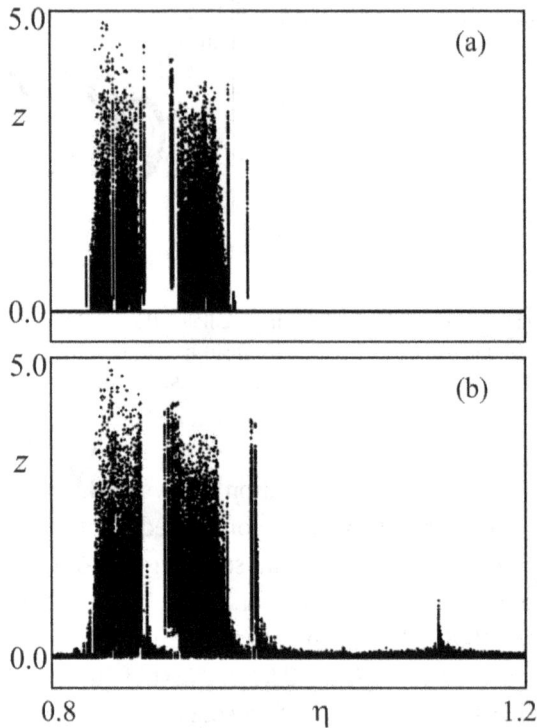

Fig. 7.27. Bifurcation diagrams of the synchronization error z corresponding to Figs. 7.26a and 7.26b, respectively, computed for (a) identical systems (7.29) — $\rho = 0$, and (b) systems with a mismatch (7.29) — $\rho = 0.03$.

In Fig. 7.27 the bifurcation graphs of the synchronization error z which correspond to the diagrams in Figs. 7.26a and 7.26b are demonstrated. The intervals of a desynchronous motion (large z) reflect the chaotic ranges from Figs. 7.26a and 7.26b. In the case of identical master and slave systems (7.30), i.e., $\rho = 0$, a distance z is equal to zero (Fig. 7.27a) in the ranges of the regular motion due to asymptotical convergence of the periodic trajectories, i.e., the CS takes place there. Obviously, the CS is impossible for non-identical systems (7.30), i.e., $\rho = 0.03$. However, we can see that the *synchronization error* remains relatively small, when the motion is regular (Fig. 7.27b), i.e., the ICS occurs. As has been mentioned in Sec. 6.4, the crucial parameter for quantifying the ICS and estimating the ELE is the ICS threshold ε. The bifurcation graph shown in Fig. 7.27b helps us to evaluate the boundary value of ε, for which the disturbed motion can be still recognized as the regular one. From our experience it results that the value of ε should be related to a magnitude of the mismatch, i.e., the ratio of ε and the maximum *synchronization error* $\sup(z)$ of uncoupled oscillators ($d = 0$) should approximate maximum variations of the mismatch. Thus, in the case under consideration we have:

$$\frac{\varepsilon}{\sup(z)} \approx \frac{\sup(\Delta\alpha(t))}{\alpha}. \tag{7.32}$$

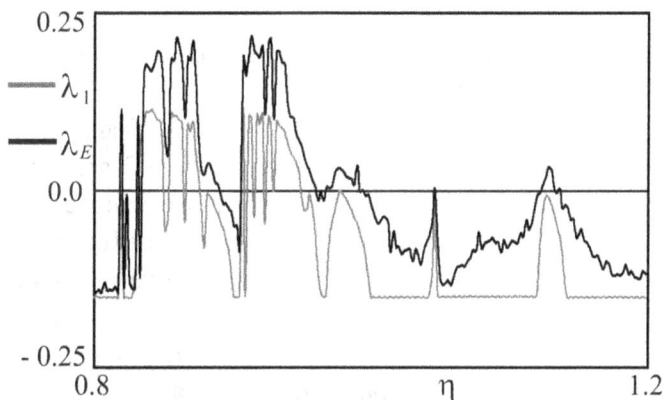

Fig. 7.28. Bifurcation plots of the LLE (in gray, $\rho = 0$) and the ELE (in black, $\rho = 0.03$) estimated from Eqs. (7.30) for the ICS threshold $\varepsilon = 0.25$, which correspond to Figs. 7.26a and 7.26b, respectively. Remaining parameters of the estimation procedure: $d_{1(0)} = -0.5$, $d_{2(0)} = 0.5$, $\delta = 0.005$, $D = 5.0$, $T_e = 5000$, $T_t = 5000$.

The plots of the LLE (in gray) and the ELE (in black) corresponding to Figs. 7.26a and 7.26b, respectively, are depicted in Fig. 7.28. They have been determined with the proposed method on the basis of Eqs. (7.30) for the undisturbed (the LLE) and disturbed (the ELE) auxiliary subsystem. It can be seen that in the chaotic ranges of bifurcation diagrams (Figs. 7.26a and 7.26b), the positive LLE and ELE are detected and *vice versa*. The intervals of the positive ELE are wider than the analogous range of the LLE and additional fields, where the ELE is larger than zero, appear in the neighborhood of the control parameter values corresponding to the period-doubling bifurcation. The comparison of the LLE and ELE plots allows us to evaluate a qualitative and quantitative influence of noise on the system dynamics.

7.4.2 *Determining the ERLE*

The outcomes of the bifurcation analysis presented in Sec. 4.3.2 have proven that the RLEs are a good tool for detection of the CS in cases of completely identical response oscillators. However, a slight (mechanical oscillators) or even infinitesimal (Henon maps) parameter mismatch of response oscillators leads to a serious disturbance of the ICS and causes a considerable shift of the synchronization threshold in the control parameter space (Figs. 4.33c and 4.35c) or even a complete destruction of the synchronous regime (Figs. 4.37c and 4.382c). Such an effective synchronization threshold is obviously of crucial significance for practical applications. A small mismatch almost does not change noticeably a value of the RLE in the considered systems. Thus, the GS between the drive and these response systems occurs in intervals of negative RLEs, despite the fact that their ICS rather cannot be observed there. The ICS takes place only for lower amplitudes of the drive (small values of the control parameter q), where the influence of damping on the dynamics of mechanical oscillators is still relatively large. An increase in the amplitude q leads to a loss of the ICS, whereas the corresponding RLE is still negative. Such a situation is a good opportunity to illustrate the idea of the ERLE.

In order to demonstrate an example of the ERLE application, let us come back to the cases presented in Sec. 4.3.2. A mismatch of the parameter α in mechanical oscillators (Eq. (4.62)) and the parameter a in

Henon maps (Eq. (4.65)) have been considered. Substituting these systems into Eqs. (6.33) and (6.34–6.35) respectively, we obtain the disturbed SRS systems in the following forms:

$$\dot{y}_1 = y_2 + d(x_1 - y_1),$$
$$\dot{y}_2 = -(\alpha + \Delta\alpha)y_1^3 - hy_2 + q[e(t) - y_1] + d(x_2 - y_2),$$

(7.33)

and

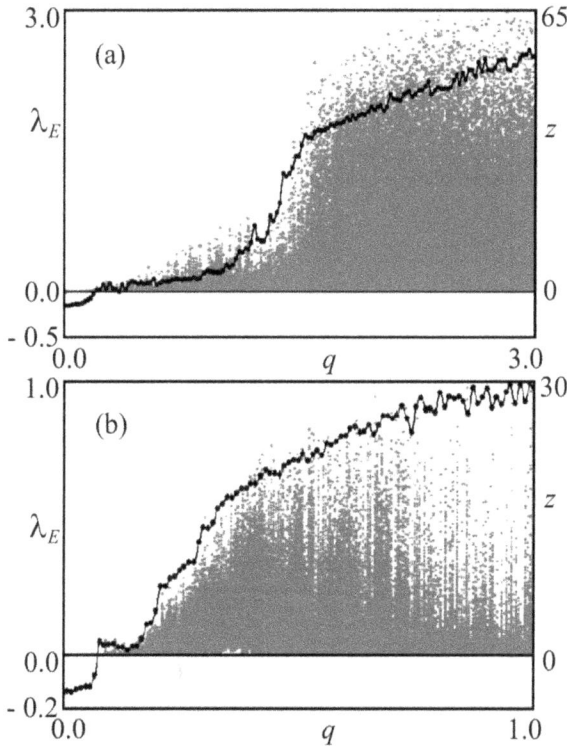

Fig. 7.29. Bifurcation diagrams of the *synchronization error* (Eq. (6.23)) (in gray) and the corresponding ERLEs estimated with the synchronization method (black line); (a) Lorenz drive (Eq. (4.63)), the ICS threshold — $\varepsilon = 0.5$, parameter mismatch — $\Delta\alpha = 0.01\alpha$; (b) continuous-time stochastic drive (Eqs. (4.64a–b)), $\varepsilon = 0.1$, $\Delta\alpha = 0.01\alpha$; other parameters of the oscillator: $\alpha = 10.0$ and $h = 0.3$. Parameters of the estimation procedure are $d_{1(0)} = -3.0$, $d_{2(0)} = 3.0$, $\delta = 0.005$, $D = 10.0$, $T_e = 5000$, $T_t = 5000$.

$$y_1^* = 1 - (a + \Delta a)(x_1 + z_1)^2 + (x_2 + z_2) + qe,$$

$$y_2^* = b(x_1 + z_1),$$
(7.34a)

$$z_1^* = [z_2 - az_1(z_1 + 2x_1) - \Delta a(x_1 + z_1)^2]\exp(-d),$$

$$z_2^* = bz_1 \exp(-d),$$
(7.34b)

where $\Delta \alpha$ and Δa represent the parameter mismatch. The related *mismatch vectors* are $[0, \Delta \alpha y_1^3]^T$ for Eq. (7.33) and $[\Delta a(x_1 + z_1)^2, 0]^T$ for Eqs. (7.34a–b).

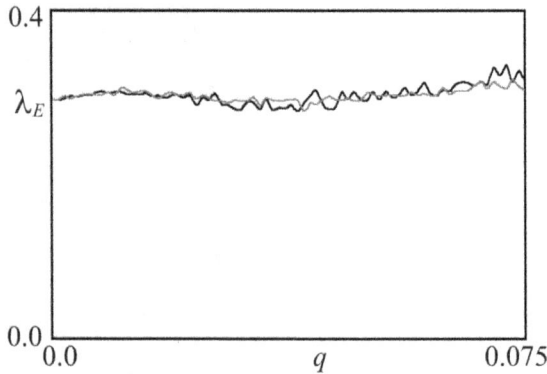

Fig. 7.30. Bifurcation diagrams of the ERLEs estimated with the synchronization method: corresponding to Figs. 4.37c — logistic map drive (Eq. (4.66), black line) and 4.38c — discrete-time random drive (Eq. (4.67), gray line); $a = 1.2$, $b = 0.3$, $\varepsilon = 0.05$, $\Delta a = 10^{-12}$. Other parameters of the estimation procedure are $d_{1(0)} = -1.0$, $d_{2(0)} = 1.0$, $\delta = 0.005$, $D = 5.0$, $n_e = 50000$, $n_t = 50000$.

In Figs. 7.29a and 7.29b the bifurcation diagrams of the *synchronization error* from Figs. 4.33c, 4.35c (slightly different Duffing oscillators, $\Delta \alpha = 0.01\alpha$) are collated with the corresponding ERLEs determined with the proposed approach (*variant 1*). We can see that the ICS state is practically stable for the negative ERLE. When the ERLE becomes positive, then the *synchronization error* grows up beyond the ε-range, and next enlarges systematically with an increasing positive ERLE. This numerical experiment confirms that the ERLEs are sufficiently precise tools to verify how robust the ICS of slightly different oscillators with the external drive is. In Fig. 7.30 the ERLE corresponding to both cases (chaotic and random excitation, Figs. 4.37c and 4.38c) of the driven Henon map with an almost infinitesimal mismatch ($\Delta a = 10^{-12}$) are depicted. These results confirm instability of the ICS between chaotic Henon systems, i.e., the ERLEs keep a relatively large positive value in the entire range of the bifurcation parameter.

In accordance with formula (7.32), in all presented examples the ICS threshold approximates value of 1% of the *synchronization error* ($\varepsilon \approx 0.01 \sup |z|$). This small parameter determines the acceptable "imperfection of the CS". It results from the above examples that the ERLE can be a good tool to distinguish between a strong and weak GS. An occurrence of the ICS indicates a strong GS, which is confirmed by the negative ERLE. On the other hand, a weak GS is sensitive to the parameter mismatch, which results in the positive ERLE (see Figs. 7.29a–b and 7.30).

7.5 Comparison of the Obtained Results

The examples presented in this Chapter show an agreement of the observed dynamical behavior of the considered systems with the sign of the largest Lyapunov exponent estimated with the proposed method. But such observations are not enough to determine a level of accuracy of the estimated λ_1 exponent value. Some ideas to judge the precision of the method can come from a comparison of its different variants.

Comparisons of the largest Lyapunov exponents of the Henon map (Eq. (6.20)) for chosen values of the parameters a and b, estimated with the synchronization method and those calculated with the classical algorithm are presented in Table 7.1 and additionally illustrated in Fig. 7.31. A very small difference between values of the exponent λ_1 in the second and third columns in Table 5.1 and the overlapping traces of the bifurcational diagrams in Fig. 5.1 prove a high accuracy of the synchronization method in relation to the mappings given by clear difference equations, both in periodic and chaotic regimes.

Table 7.1. Comparison of the values of the LLE of the Henon map calculated with the classical algorithm and the proposed method.

Parameters		Classical algorithm	Synchronization method
$a=1.40$,	$b=0.30$	+0.4186	+0.4050
$a=1.10$,	$b=0.30$	+0.1811	+0.1830
$a=0.70$,	$b=0.30$	−0.6019	−0.6020
$a=0.30$,	$b=0.30$	−0.0785	−0.0770

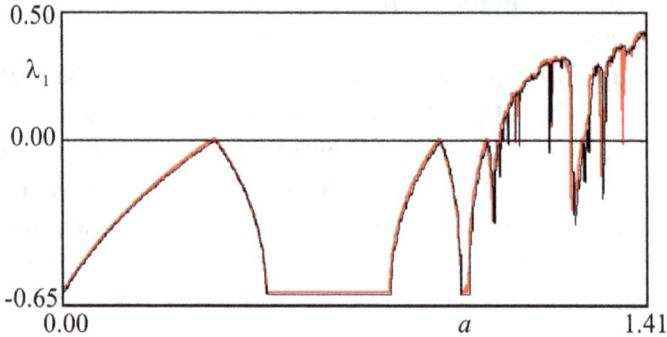

Fig. 7.31. Comparison of the bifurcational diagrams of the LLE of the Henon map (Eq. (6.20)) as a function of the parameter a, obtained with the synchronization method (red line) and with the classical algorithm (black line); $b=0.30$; numerical procedure parameters: $\varepsilon=0.001$, $D=1.00$, $n_f=5000$, $n_e=5000$, $d_{1(0)}=-1.00$, $d_{2(0)}=1.00$, $\delta d=0.001$.

To verify the accuracy of the synchronization method in the case of impact (Eq. (7.1a–b)) and stick-slip (Eq. (7.7)) oscillators, the Müller method of calculation of Lyapunov exponents for dynamical systems with discontinuities (Müller (1995)), which is described in Chapter 5, has been applied. On the other hand, for a comparison of the outcomes obtained with different variants of the synchronization method, the following relations arising from formula (5.31) have been used:

$$\lambda_1 \approx \frac{\lambda_P}{T} \approx \frac{\lambda_I}{\tau_A} \qquad (7.35)$$

in the case of an impact oscillator, where $T=2\pi/\eta$ is the excitation period, and $\tau_A=\Sigma_n\tau_n/n$ is an average time between impacts.

Table 7.2. Comparison of the largest Lyapunov exponent λ_1 of the mechanical oscillator with impacts (Eq. (7.1a–b)); $\alpha=1.00$, $h=0.00$, $p=1.00$, $R=0.80$, $\delta_0=0.00$.

η	T	τ_A	Müller method λ_1	Variant 1 λ_1	*Variant 3*			
					Impact map		Poincaré map	
					λ_I	λ_I/τ_A	λ_P	λ_P/T
3.00	2.094	2.094	–0.1065	–0.1050	–0.2230	–0.1065	–0.2240	–0.1070
3.105	2.024	1.712	0.1172	0.0980	0.2030	0.1186	0.2380	0.1176
3.121	2.013	1.611	–0.1381	–0.1220	–0.2230	–0.1384	–0.2790	–0.1386
3.126	2.010	1.887	0.1071	0.0950	0.2050	0.1086	0.2270	0.1129

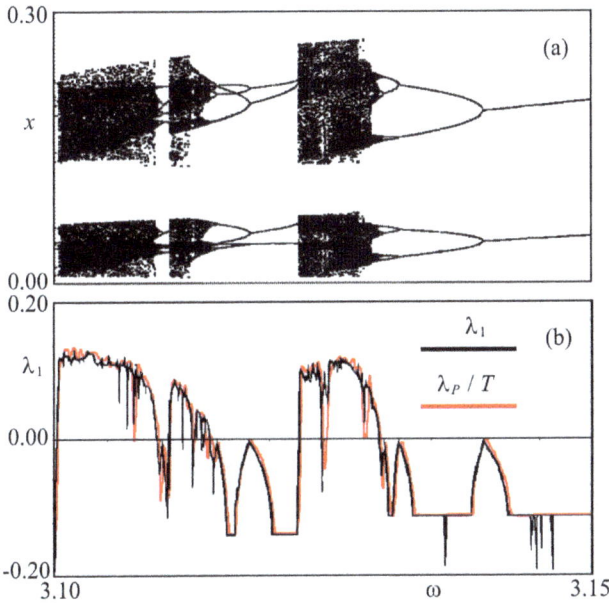

Fig. 7.32. Bifurcational diagram of the mechanical oscillator with impacts (Eq. (7.1a–b)) as a function of the excitation frequaency ω (a), and a comparison of the related LLE (b), obtained with the earier algorithm (in black) and the synchronization method (*variant 3* based on the Poincaré map — in red); $\alpha=1.00$, $h=0.00$, $p=1.00$, $R=0.80$, $\delta_0=0.00$. Numerical procedure parameters: $\varepsilon=0.001$, $D=1.00$, $n_t=500$, $n_e=500$, $d_{1(0)}=-0.50$, $d_{2(0)}=0.50$, $\delta d=0.005$.

A comparison of the calculated values of the exponent λ_1 with their counterparts estimated with *variant* 1 and *variant* 3 of the synchronization method, for chosen values of the impact system parameters, is presented in Table 7.2. In Fig. 7.32a, a bifurcational diagram of system (7.1a–b) and the corresponding LLE in Fig. 7.32b are shown. The Müller algorithm result is shown in black, while the synchronization method is presented in red. *Variant* 3 was based on the Poincaré maps. Those results, shown in Table 7.2 and Fig. 7.32b, prove a high precision of the method again, especially in the periodic regimes. In the chaotic regimes, differences in both methods for the λ_1 values are not larger than 10%.

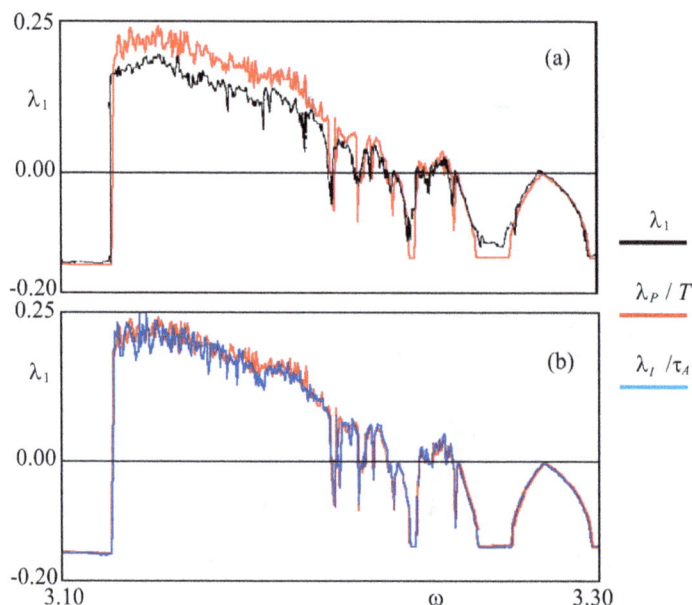

Fig. 7.33. Comparison of the bifurcational diagrams of the LLE (a comparison of the use of formula (7.35)) presented in Figs. 7.6b and 7.6c (a), and 7.6c and 7.6d (b).

The results obtained with the synchronization method only are shown in Figs. 7.33a and 7.33b, which are related to those presented in Figs.7.6b–d, obtained with formula (7.35). In the case of *variant* 1 of the method, differences of the compared λ_1 are slightly larger — see Table 7.2, which can be a result of a lower accuracy of this variant in the system with discontinuities. This is also proven by the diagrams in Figs. 7.33a and 7.33b, where one can spot only small differences in both cases of *variant 3* — Fig. 7.33b, while larger differences occur between *variants* 1 and 3 — Fig. 7.33a.

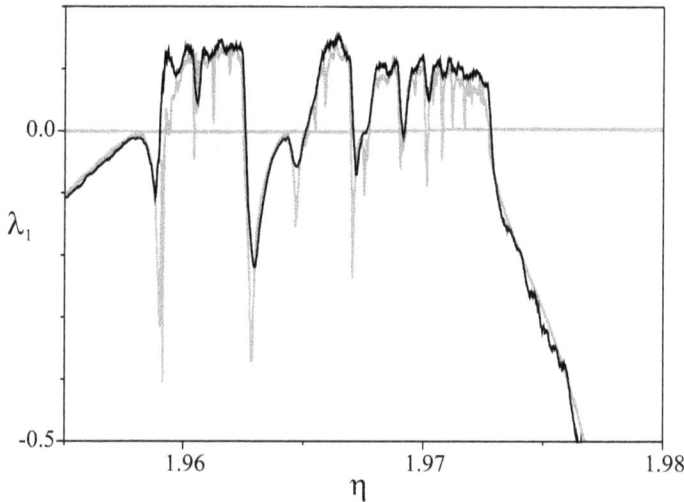

Fig. 7.34. LLE calculated with the synchronization method (black) compared with the Müller (1995) algorithm (gray) for the stick-slip oscillator (Eq. (7.7)). Numerical procedure parameters: $\varepsilon=0.001$, $D=1.00$, $n_r=500$, $n_e=500$, $d_{1(0)}=-0.50$, $d_{2(0)}=0.50$, $\delta d=0.005$.

In the case of a stick-slip oscillator (Eq. (7.7)), the proposed method has been also verified by a comparison with the Müller algorithm. In Fig. 7.34, such a comparison of the LLEs is presented. In order to simplify a comparison of the results, the parameters of the friction oscillator are the same as those used by in Ref. Müller (1995). We can see that there exists some small discrepancy between the courses of the LLEs. However, the most important is that they both cover the same ranges of the control parameter, where the LLE is positive or negative. Thus, the results obtained with the proposed approach approximate the LLE with a good precision.

A comparison of the diagrams of the LLE obtained with different variants of the method in the Van der Pol system with time delay is illustrated in Fig. 7.35, which is another comparison of Figs. 7.23b and 7.23c. It has been obtained on the basis of the relation emerging from relation (5.31) with the results estimated with synchronization according to *variant 1* (λ_1) and *variant 3* on the basis of the τ–map (λ_τ), which in this case is as follows:

$$\lambda_1 \approx \frac{\lambda_{1\tau}}{\tau_0} . \qquad (7.36)$$

In Fig. 7.35 one can observe that relation (7.36) is fulfilled in a wide range of the control parameter κ_0. These facts also confirm the efficiency of the proposed method with respect to systems with time delay, no matter which variant is used.

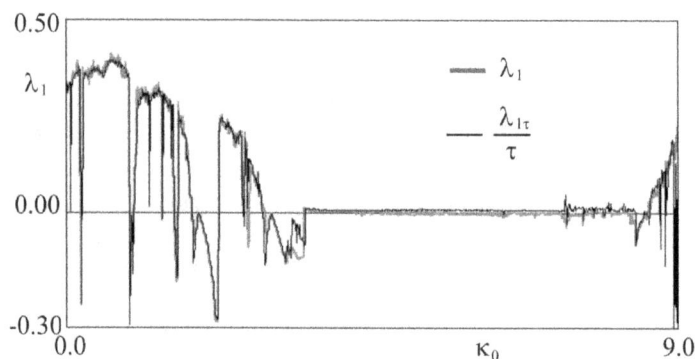

Fig. 7.35. Comparison of the bifurcational diagrams of the largest Lyapunov exponents using formula (7.36) presented in Figs. 7.23b in black, and 7.23c, in gray.

A verification of the RLEs estimated by means of the synchronization method are shown in Fig. 4.33a, which illustrates how the corresponding largest RLEs, calculated by means of the classical algorithm (in black) and the ones estimated with the synchronization method (in gray) vary. It is clearly visible that their courses are mutually overlapped. This fact confirms the precision of determination of the maximum RLE with the proposed approach.

Bibliography

Abarbanel, H. D. I., Rulkov, H. F., Sushchik, M. M. (1996). Generalized synchronization of chaos: The auxiliary system approach, *Phys. Rev.* E, 53, pp. 4528–4535.

Afraimovich, V. S., Cordonet, A., Rulkov, H. F. (2002). Generalized synchronization of chaos in noninvertible maps, *Phys. Rev.* E, 66, 016208, 6p.

Albert, R., Jeong, H. and Barabási, A.-L. (1999). The diameter of the world-wide web, Nature (London), 401, pp. 130–131.

Anishchenko V. C. (1990). Complex oscillations in simple systems, Moscow: Nauka Publisher, 312 pp. (in Russian).

Ashwin, P., Buescu, J., and Stewart, I. (1994). Bubbling of attractors and synchronization of chaotic oscillators, *Phys. Lett.* A, 193, pp. 126–139.

Atay, F. M. and Bıyıkoğlu, T. (2005). Graph operations and synchronization of complex networks, *Phys. Rev.* E, 72, 016217, 7p.

Awrejcewicz, J. and Wojewoda J. (1989). Observation of chaos in a nonlinear oscillator with delay: Numerical study, *KSME Journal*, 3(1), pp. 15–24.

Barahona, M. and Pecora, L. M. (2002). Synchronization in small-world systems, *Phys. Rev. Lett.*, 89, 054101 4p.

Bar–Eli, K. (1985). On the stability of coupled chemical oscillators, *Physica* D 14, pp. 242–252.

Barnett, S. and Storey, C. (1970) Matrix Methods in Stability Theory, Thomas Nelson and Sons Ltd., Suffolk, 148p.

Belykh, V. N., Belykh, I. V., Hasler, M. (2004). Connection graph stability method for synchronized coupled chaotic systems, *Physica* D, 195, pp. 159–187.

Belykh, I. V., Belykh, V. N., Nevidin, K. V., Hasler, M. (2003). Persistent clusters in lattices of coupled nonidentical chaotic systems, *Chaos*, 13, pp. 165–178.

Benettin, G., Galgani, L., Strelcyn, J. M. (1976). Kolmogorov entropy and numerical experiment, *Phys. Rev.* A, 14, pp. 2338–2345.

Benettin, G., Galgani, L., Giorgilli, A., Strelcyn, J.M. (1980). Lyapunov exponents for smooth dynamical systems and Hamiltonian systems; a method for computing all of them, Part I: Theory, *Meccanica*, 15; pp. 9–20.

Benettin, G., Galgani, L., Giorgilli, A., Strelcyn, J.M. (1980). Lyapunov exponents for smooth dynamical systems and Hamiltonian systems; a method for computing all of them, Part II: Numerical application. *Meccanica*, 15; pp. 21–30.

Biggs, N. (1993). Algebraic Graph Theory. Cambridge Mathematical Library, 2nd ed., 216p.

Birkhoff, G. D. (1927). Dynamical Systems, AMS <u>Colloquium Publications</u>, Providence, 305p (J. Moser reprint).

Blazejczyk–Okolewska, B. (1995). Analysis of irregular motion in impact oscillator, PhD thesis, Technical University of Lodz, 127p. (in Polish).

Blazejczyk–Okolewska, B., Kapitaniak, T. (1996). Dynamics of impact oscillator with dry friction, *Chaos Solitons Fractals*, 7(9), pp. 1455–1459.

Blekhman, I. (1988). Synchronization in Science and Technology, ASME Press, New York, 350p.

Boccaletti, S., Kurths, J., Osipov, G., Valladares, D.L., Zhou, C.S. (2002). The synchronization of chaotic systems, *Phys. Rep.*, 366, pp. 1–101.

Brown, R., Rulkov, N. F. (1997). Designing coupling that guarantees synchronization between identical chaotic systems, *Phys. Rev. Lett.*, 78, pp. 4189–4192.

Cao, L–Y. and Lai,Y–Ch. (1998). Antiphase synchronism in chaotic systems, *Phys. Rev. E*, 58(1), pp. 382–386.

Chen, G. and Duan, Z. (2008). Network synchronizability analysis: A graph-theoretic approach, *Chaos*, 18(3), 037102.

Chen, Y., Rangarajan, G., Ding, M. (2003). General stability analysis of synchronized dynamics in coupled systems, *Phys. Rev. E*, 67, 026209, 4p.

Chua L.O., Komuro M. & Matsumoto T. (1986). The double scroll family parts I and II. IEEE *Trans. Circuits Syst.*, 33(11), pp. 1073–1118.

Chua, L.O. (1993). Global unfolding of Chua's circuit. *IEICE Trans. Fund. Elec.*, E76–A(5), pp. 704–734.

Chua, L. O. (1994). Chua's circuit an overview ten years later, *J. Circuit Syst. Comput.*, 4(2), pp. 117–159.

Cuomo, K., Oppenheim, A. and Strogatz, S. (1993). *IEEE Trans. Circuits Syst.* II, 40, pp. 626–633.

Dabrowski, A. (2002). Dynamics of vibration absorber with limiters of motion amplitude, PhD thesis, Technical University of Lodz, 84p. (in Polish).

Den Hartog, J. P. (1934). Mechanical Vibrations, Doveer Publ., New York, 436p.

Dmitriev, A. S., Shirokov, M. and Starkov, S. O. (1997). Chaotic synchronization in ensembles of coupled maps, *IEEE Trans. Circuits Syst. I. Fund. Th. Appl.*, 44(10), 918–926.

Duan, Z., Chen, G. and Huang, L. (2007). Complex network synchronizability: Analysis and control, *Phys. Rev. E*, 76, 056103, 6p.

Duan, Z., Liu, C. and Chen, G. (2008). Network synchronizability analysis: The theory of subgraphs and complementary graphs. *Physica D: Nonlinear Phenomena*, 237 (7), pp. 1006–1012.

Eckmann, J.P., Kamphorst, S.O., Ruelle, D., Ciliberto, S. (1986). Lyapunov exponents from a time series. *Phys. Rev. Lett.*, 34(9), pp. 4971–4979.

Femat, R., Kocarev, L., van Gerven, L., Monsivais–Perez, M. E. (2005). Towards generalized synchronization of strictly different chaotic systems, *Phys. Lett.* A, 342, pp. 247–255.

Fink, K., Johnson, G., Carroll, T. L., Mar, D., and Pecora, L. M. (2000). Three coupled oscillators as a universal probe of synchronization stability in coupled oscillator arrays, *Phys. Rev.* E, 61, pp. 5080–5090.

Fujisaka, H., Yamada, T. (1983a). Stability theory of synchronized motion in coupled-oscillator systems. *Prog. Theor. Phys.*, 69(1), pp. 32–47.

Fujisaka, H., Yamada, T. (1983b). Stability theory of synchronized motion in coupled-oscillator systems II: The mapping approach, *Prog. Theor. Phys.* 70, 1240–1248.

Gade P. M., Cerdeira H. & Ramaswamy R., (1995). Coupled maps on trees, *Phys. Rev.* E, 52, 2478–2485.

Gade, P. M. (1996). Synchronization in coupled map lattices with random non-local connectivity, *Phys. Rev.* E, 54, pp. 64–70.

Galvanetto, U. (2000). Numerical computation of Lyapunov exponents in discontinuous maps implicitly defined, *Com. Phys. Comm.*, 131, pp. 1–9.

Gauthier, D. J. and Bienfang, J. C. (1996). Intermittent loss of synchronization in coupled chaotic oscillators: Toward a new criterion for high-quality synchronization, *Phys. Rev. Lett.*, 77, pp. 1751–1754.

Hale, J. (1997). Diffusive Coupling, Dissipation, and Synchronization, *J. Dyn. Diff. Eq.*, 9(1), pp. 1–52.

Heagy, J. F., Carroll, T. L. and Pecora L. M. (1994). Synchronous chaos in coupled oscillators systems, *Phys. Rev.* E, 50(3), pp. 1874–1885.

Henon, M., Heiles, C. (1964). The applicability f the third integral of the motion: Some numerical results, *Astronomic Journal*, 69, pp. 73–79.

Henon, M.(1976). A two dimensional map with a strange attractor, *Commun. Math. Phys.*, 50, pp. 69–77.

Hertz, J., Krogh, A. and Palmer, R. (1991). Introduction to the Theory of Neural Computation, Addison–Wesley, Reading, MA, 327p.

Hilborn, R.C.(1994). Chaos and Nonlinear Dynamics, Oxford University Press, New York, 672p.

Hinrichs, N., Oestreich, M., Popp, K. (1997). Dynamics of oscillators with impact and friction, *Chaos Solitons Fractals*, 4(8), pp. 535–558.

Huygens, C. (1673). Horologium Oscilatorium, Paristis.

Jin, L., Lu, Q. S., Twizell, E. H. (2006). A method for calculating a spectrum of Lyapunov exponents by local maps in non-smooth impact-vibrating systems, *J. Sound Vib.*, 298(4–5), pp. 1019–1033.

Kaneko, K. (1994). Relevance of Clustering to Biological Networks, *Physica* D, 75, pp. 55–73.

Kantz, H. (1994). A robust method to estimate the maximal Lyapunov exponent of a time series. *Phys. Lett.* A, 185(1), pp. 77–87.

Kapitaniak, T., Sekieta, M., Ogorzalek, M. (1996). Monotone synchronization of chaos. *Int. J. Bifurcation Chaos*, 6, pp. 211–215.

Kaplan, J.; Yorke, J. (1979). Chaotic behavior of multidimensional difference equations. In Functional Differential Equations and Approximation of Fixed Points, Peitgen, H.O.; Walther, H.O.; eds.; Springer–Verlag: New York, pp. 228–237.

Kocarev, L., Parlitz, U. (1995). General Approach for Chaotic Synchronization with Applications to Communication, *Phys. Rev. Lett.*, 74, pp. 5028–5031.

Kocarev, L., Parlitz, U. (1996). Generalized synchronization, predictability, and equivalence of unidirectionally coupled dynamical systems, *Phys. Rev. Lett.*, 74, pp. 1816–1819.

Kuznetsov, S. P. (2001). Dynamical Chaos, Fizmatlit, Moscow, 296p.

Liu, Y., Rios Leite, J.R. (1994). Coupling of two chaotic lasers, *Phys. Lett.* A, 191, pp. 134–138.

Lorenz, E.N. (1963). Deterministic nonperiodic flow. *J. Atmospheric Sciences*, 20(2), pp. 130–141.

Lyapunov, A.M. (1947). Probleme General de la Stabilite du Mouvment. Annales Mathematical Study, **17**, Princeton University Press, Princeton, New Jersey, 474p.

Mandelbrot, B. B. (1982). The Fractal Geometry of Nature. Freeman, San Francisco, 460p.

Massoler, C. (2001). Anticipation in the Synchronization of Chaotic Semiconductor Lasers with Optical Feedback, *Phys. Rev. Lett.*, 86, pp. 2782–2785.

Müller, P. C. (1995). Calculation of Lyapunov exponents for dynamical systems with discontinuities. *Chaos Solitons Fractals*, 5(9), pp. 1671–1681.

Nishikawa, T., Motter, A. E., Lai, A. E. and Hoppensteadt, F. C. (2003). Heterogeneity in oscillator networks: Are smaller worlds easier to synchronize? *Phys. Rev. Lett.*, 91(1), 014101, 4p.

Nusse, H. E. and Yorke, J. A. (1994). *Dynamics: Numerical Explorations*, (Springer–Verlag, New York, 558p.

Oestreich, M., Hinrichs, N., Popp, K.(1996). Bifurcation and stability analysis for a non-smooth friction oscillator, *Arch. Appl. Mech.*, 66, pp. 301–314.

Oestreich, M. (1998). Analysis of non-smooth vibrations, PhD thesis, VDI 11 No. 258, VDI Verlag GmbH, Düsseldorf, (in German).

Oseledec, V.I. (1968). A multiplicative ergodic theorem: Lyapunov characteristic numbers for dynamical systems, *Trans. Moscow Math. Soc.*, 19, pp. 197–231.

Park, E.-H., Feng Z. and Durand D. M. (2008). Diffusive coupling and network periodicity: a computational study, *Biophys J.*, 95(3), pp. 1126–1137.

Parker, T.S. and Chua, L.O. (1989). Practical Numerical Algorithms for Chaotic Systems, Springer–Verlag, Berlin, 181p.

Parlitz, U. (1992). Identification of true and spurious Lyapunov exponents from time series, *Int. J. Bifurcation Chaos*, 2(1), pp. 155–165.

Pikovsky, A.S. (1984). On the interaction of strange attractors, *Zeitschrift Physik* B, 55, pp. 149–155.

Pikovsky, A.S., Rosenblum, M.G. and Kurths, J. (1997). Phase synchronization in drive and coupled chaotic oscillators, *IEEE Trans. Circuits Syst.*, 44(10), pp. 874–881.

Pikovsky, A. S., Rosenblum, M. G. and Kurths. J. (2001). Synchronization. A Universal Concept in Nonlinear Sciences, Cambridge University Press, Springer, Berlin. 411p.

Pecora, L.M. and Carroll, T.L. (1990). Synchronization in chaotic systems, *Phys. Rev. Lett.*, 64, pp. 821–824.

Pecora, L.M., and Carroll, T.L. (1991). Driving systems with chaotic signals, *Phys. Rev. A*, 44, pp. 2374–2383.

Pecora, L.M. and Carroll, T.L. (1998). Master stability functions for synchronized coupled systems, *Phys. Rev. Lett.* 80(10), pp. 2109–2112.

Pecora, L.M. (1998). Synchronization conditions and desynchronizing patterns in coupled limit-cycle and chaotic systems, *Phys. Rev. E*, 58(1), pp. 347–360.

Pecora, L. M., Carroll, T. L., Johnson, G., Mar, D., Fink, K. (2000). Synchronization stability in coupled oscillator arrays: solution for arbitrary configurations, *Int. J. Bifurcation Chaos*, 10(2), pp. 273–290.

Perlikowski, P. (2007). Complete synchronization in networks of coupled nonlinear systems, PhD thesis, Technical University of Lodz, 112p. (in Polish).

Perlikowski, P., Jagiello, B., Stefanski, A. and Kapitaniak, T. (2008). Experimental observation of ragged synchronizability, *Phys. Rev. E*, 78(1), 017203, 4p.

Platt, N., Spiegel, A. and Tresser, C. (1993). On-off intermittency: A mechanism of bursting, *Phys. Rev. Lett.*, 70, pp. 279–282.

Poincaré, H. (1913). The Fundation of Science: Science and Method, The Science Press, Lancaster PA (English translation 1946), 397p.

Pogromsky, A. Yu., Santoboni, G., Nijmeijer, H. (2002). Partial synchronization: from symmetry towards stability, *Physica* D, 172, pp. 65–87.

Popp, K., Hinrichs, N., Oestreich, M. (1996). Analysis of a self-excited friction oscillator with external excitation, *Dynamics with Friction: Modeling, Analysis and Experiment* (Guran A., Pfeiffer F., Popp K. eds.), World Scientific Publishing, Singapore.

Press, W.H., Teukolsky, S.A., Vetterling, W.T. and Flannery, B.P., Numerical Recipes in C++, Cambridge University Press, 1002p.

Pyragas, K. (1996). Weak and strong synchronization of chaos, *Phys. Rev. E* **54**, pp. 4508–4511.

Pyragas, K. (1998). Properties of generalized synchronization of chaos, *Nonlinear Analysis: Modelling and Control*, 3, Vilnius, IMI, 29p.

Rosenblum, M.G., Pikovsky, A.S., and Kurths, J. (1996). Phase synchronization of chaotic oscillators, *Phys. Rev. Lett.*, 76(11), pp. 821–824.

Rosenblum, M.G., Pikovsky, A.S. and Kurths, J. (1997). From phase to lag synchronization in coupled chaotic oscillators, *Phys. Rev. Lett.*, 78(22), pp. 4193–4196.

Rosenstein, M.T., Collins, J.J., De Luca, C.J. (1993). A practical method for calculating largest Lyapunov exponents from small data sets, *Physica* D, 65(1–2), pp. 117–134.

Rössler, O. E. (1976). An equation for continuous chaos, *Phys. Lett.* A, 57, pp. 397–398.

Rulkov, F., Sushchik, M. M., Tsimring, L. S., Abarbanel, H. D. I. (1995). Generalized synchronization of chaos in directionally coupled systems, *Phys. Rev.* E, 51, pp. 980–984.

Rulkov, F., Sushchik, M. M. (1997). Robustness of synchronized chaotic oscillations, *Int. J. Bifurcation Chaos*, 7, pp. 625–643.

Sano M., Sawada Y.: Measurement of the Lyapunov spectrum from a chaotic time series. Physical Review Letters, **55**, 1985, 1082–1085.

Schuster, H.G. and Just W. (2006). Deterministic Chaos: An Introduction. Wiley–VCH (4th enlarged ed.), 312p.

Sekieta, M. and Kapitaniak T. (1996). Practical synchronization of chaos via nonlinear feedback scheme, *Int. J. Bifurcation Chaos* 6 (10), pp. 1901–1907.

Shimada, I. and Nagashima, T. (1979). A numerical approach to ergodic problem of dissipative dynamical systems, *Prog. Theor. Phys.*, 61(6), pp. 1605–1616.

Rayleigh, J. (1945). Theory of Sound, Dover Publishing, 480p.

Shinozuka, M. and Deodatis, G. (1992). Simulation of stochastic processes by spectral representation, *Appl. Mech. Rev.*, **44**, 191–204.

Singh, A., Joseph, D.D. (1989). Autoregressive methods for chaos on binary sequences for the Lorenz atrractor, *Phys. Lett.* A, 135, pp. 247–251.

Soen, Y., Cohen, N., Lipson, D. and Braun, E. (1999). Emergence of spontaneous rhythm disorders in self-assembled networks of heart cells. *Phys. Rev. Lett.* 82, 3556–3559.

Souza, S.L.T. and Caldas, I.L. (2004). Calculation of Lyapunov exponents in systems with impacts, *Chaos Solitons Fractals*, 19, pp. 569–579.

Stefański, A., Kapitaniak, T. (2000). Using chaos synchronization to estimate the largest Lyapunov exponent of non-smooth systems, *Discrete Dyn. Nat. Soc.*, 4, pp. 207–215.

Stefański, A. (2000). Estimation of the largest Lyapunov exponent in systems with impacts, *Chaos Solitons Fractals*, 11 (15), pp. 2443–2451.

Stefański A., Kapitaniak T. (2003a). Synchronization of mechanical systems driven by chaotic or random excitation, *J. Sound Vib.*, 260, pp. 565–576,

Stefański, A., Kapitaniak, T. (2003b). Estimation of the dominant Lyapunov exponent of non-smooth systems on the basis of maps synchronization, *Chaos Solitons Fractals*, 15, pp. 233–244.

Stefański, A., Kapitaniak, T. (2003c). Synchronization of two chaotic oscillators via negative feedback mechanism, *Int. J. Solids Structures*, 40, pp. 5175–5185.

Stefański, A., Wojewoda, J., Kapitaniak, T., Yanchuk, S. (2004). Simple estimation of synchronization threshold in ensembles of diffusively coupled chaotic systems, *Phys. Rev. E*, 70, 026217, 11p.

Stefanski, A. (2008). Quantifying the synchronizability of externally driven oscillators, *Chaos*, 18, 013106, 13p.

Stoop, R., Meier, P.F. (1988). Evaluation of Lyapunov exponents and scaling functions from time series, *J. Opt. Soc. Am.* B, 5(5), pp. 1037–1045.

Takens, F. (1981). Detecting strange attractors in turbulence. Lecture notes in mathematics, 898, pp. 366.

Terry, J.R., Thornburg, K.S. Jr., De Shazer, D.J., Vanwiggeren, G.D., Zhu, S., Ashwin, P., Roy, R. (1999). Synchronization of Chaos in an Array of Three Lasers, *Phys. Rev. E*, 59, pp. 4036-4043.

Timoshenko, S., Young, D. H. (1928). Vibration Problems in Engineering, D. Van Nostrand Company, Inc. Princeton, New York, 357p.

Van der Pol, B. (1920). Theory of the amplitude of free forced triod vibration, *Radio Rev.*, 1, pp. 701–710.

Van der Pol, B. (1927). Forced Oscillationsin a Circuit with non-linear Resistance, *Phil. Mag.*, 3, pp. 65–80.

Voss, H.U. (2000). Anticipating chaotic synchronization, *Phys. Rev. E*, 61, pp. 5115–5119.

Voss, H.U. (2001). Dynamic Long-Term Anticipation of Chaotic States, *Phys. Rev. Lett.*, 87, 014102, 4p.

Watts, D. J., Strogatz, S. H. (1998). Collective dynamics of 'small-world' networks, *Nature* (London), 393, pp. 440–442.

Watts, D. J., (1999). Small Worlds: The Dynamics of Networks Between Order and Randomness, Princeton University Press, Princeton, 262p.

Winful, H.G. and Rahman, L. (1990). Synchronized chaos and spatiotemporal chaos in arrays of coupled lasers, *Phys. Rev. Lett.*, 65, pp. 1575–1578.

Wolf, A., Swift, J. B., Swinney H. L., Vastano, J. A. (1985). Determining Lyapunov exponents from a time series, *Physica* D, 16, pp. 285–317.

Wolf, A. (1986). Quantifying chaos with Lyapunov exponents, Chaos (V. Holden eds.), Manchester University Press, Manchester, pp. 273–290.

Wu, C. W. and Chua, L. O. (1994). A unified framework for synchronization and control of dynamical systems, *Int. J. Bifurcation Chaos*, 4(4), pp. 979–998.

Wu, C. W. and Chua, L. O. (1995a). Synchronization in an Array of Linearly Coupled Dynamical Systems, IEEE Trans. Circuits Syst. 42(8), pp. 430–447.

Wu, C. W. and Chua, L. O. (1995b). Application of graph theory to the synchronization in an array of coupled nonlinear oscillators, *IEEE Trans. Circuits Syst.*, 42(8), pp. 494–497.

Wu, C. W. and Chua, L. O. (1996). On a conjecture regarding the synchronization in an array of linearly coupled dynamical systems, *IEEE Trans. Circuits Syst.*, 43, pp. 161–165.

Wu, C. W. (2002). Simple three-oscillator universal probe for determining synchronization stability in coupled arrays of oscillators, *Int. J. Bifurcation Chaos*, 12(10), pp. 2233–2238.

Wu, C. W. (2005). Synchronizability of networks of chaotic systems coupled via a graph with a prescribed degree sequence, *Phys. Lett.* A, 346, pp. 281–287.

Yanchuk, S., Maistrenko, Y. and Mosekilde, E. (2003). Synchronization of continuous-time chaotic oscillators, *Chaos*, 13(1), pp. 388–400.

Zaks, M. A., Park, E.–H., Rosenblum, M. G., Kurths, J. (1999). Alternating Locking Ratios in Imperfect Phase Synchronization, *Phys. Rev. Lett.*, 82, pp. 4228–4231.

Index

www.ingramcontent.com/pod-product-compliance
Lightning Source LLC
Chambersburg PA
CBHW050559190326
41458CB00007B/2108

* 9 7 8 9 8 1 2 8 3 7 6 6 0 *